CHANGING CLIMATE CHANGING LIVES

Against the Backdrop of Past Changes

David R. Johnson

outskirts press

Changing Climate Changing Lives
Against the Backdrop of Past Changes
All Rights Reserved.
Copyright © 2020 David R. Johnson
v1.0

The opinions expressed in this manuscript are solely the opinions of the author and do not represent the opinions or thoughts of the publisher. The author has represented and warranted full ownership and/or legal right to publish all the materials in this book.

This book may not be reproduced, transmitted, or stored in whole or in part by any means, including graphic, electronic, or mechanical without the express written consent of the publisher except in the case of brief quotations embodied in critical articles and reviews.

Outskirts Press, Inc.
http://www.outskirtspress.com

PB ISBN: 978-1-9772-2672-3
HB ISBN: 978-1-9772-2728-7

Cover Photo © 2020 www.gettyimages.com. All rights reserved - used with permission.

Outskirts Press and the "OP" logo are trademarks belonging to Outskirts Press, Inc.

PRINTED IN THE UNITED STATES OF AMERICA

CONTENTS

	Preface	i
Chapter One	Departure	1
Chapter Two	Numbers	5
Chapter Three	Measurements	27
Chapter Four	Calendars	63
Chapter Five	Clocks	82
Chapter Six	Temperature	115
Chapter Seven	Climate	141
Chapter Eight	Arrival	171
Appendix	Science – An International Enterprise	181
	Abbreviations	186
	Notes	188
	Bibliography	218
	Illustrations	241
	Acknowledgments	245

PREFACE

I have been stirred to write this book by my deepening concern about our planet's climate. Approaching age 80, my concern runs to our children, grandchildren and future generations – those in my own family, for sure, but those in other families around the globe as well. The attention I have been giving to climate change has driven me to examine other life-changing challenges which earlier generations have faced. That means history is my launching point.

Chapter One sets the stage. A nonstop passenger flight from Washington to Beijing departs and then, tracking this flight to its destination, Chapters Two through Six deal with the histories of numbers, measurements, calendars, clocks and temperature. Each is itself an interesting tale about humans finding ways to improve their own lives and, by doing so, to achieve ever closer ties with others around the globe. Then Chapter Seven takes our focus to the climate – today's and tomorrow's.

As years sweep by, the alarms coming from those who monitor the global climate are increasingly frequent and increasingly urgent. Both the number of scientists committing their careers to our climate and the portion of those scientists who are sounding alarms have been growing. Meanwhile, the number of political office holders giving attention to the alarms has been increasing, yes – but at a much slower rate, and beyond that some office holders are still quite willing to be identified as deniers of climate change. Alongside the scientists and the politicians, the rest of us are gradually becoming more aware of climate change, but our collective awareness has not yet pushed most of us to make many compensating changes in our own lifestyles. In short, the gap is growing between scientific observations about climate change and political and individual commitments to deal with that change.

The problem may well be that we are too closely attached to our own places on this planet to see it as a whole. That is not the case, however, for the very few who have been chosen to staff the international space station and other human ventures into space. One of those fortunate few is the German astronaut Alexander Gerst (1976 –), who has shared with us his 2014 observation from space:

> *"Some things that on Earth we see in the news every day and thus almost tend to accept as a 'given,' appear very different from our perspective. We do not see any borders from space. We just see a unique planet with a thin, fragile atmosphere, suspended in a vast and hostile darkness. From up here it is crystal clear that on Earth we are one humanity, we eventually all share the same fate."* [1]

The view of Earth from space, seeing ourselves as one humanity, without borders that divide us, is one for all of us to plant firmly in our own minds.

Climate change has been tugging to get our attention since the middle of the 20th century. But as Chapters Two through Six point out, for centuries long before the past half-century our ancestors around the globe confronted their own challenges. My perspective is that by looking backward through history at other fundamental challenges which our global ancestors confronted we can now, in our own lifetimes in the 21st century, attach to our growing awareness of climate change the much-needed individual commitments to deal with it.

Chapter Eight, the final one, offers perspectives about how we should be reacting to climate change. We should all be asking ourselves about our willingness to make changes in our own lives which are both sufficient enough and soon enough to respond meaningfully to the alarms. On a global basis, will we learn how to minimize the effects of global climate change while at the same time adapting our lifestyles to whatever change inevitably comes? One thing is clear: the global climate makes all of us interdependent because we are all inhabitants of a single planet. As the German astronaut has told us, "… on Earth we are one humanity, we eventually all share the same fate." Our best choice, as one global population, is to share information about both minimizing the effects and adapting to the consequences of climate change.

David R. Johnson

Chapter One
DEPARTURE

Unify Airlines Flight 1230 departed Washington-Dulles on February 16, on time at 2:15 P.M., for its daily nonstop service to Beijing. Its scheduled flight time was 14 hours, 15 minutes, which would put it on the ground in Beijing at 5:30 P.M., local time, on February 17. Total distance to be traveled was 6,915 miles.

The plane carried 233 passengers, including 112 Americans, 76 Chinese, 22 Canadians, 14 Latin Americans and 9 Europeans. In addition, on board were two cockpit crews (each a pilot plus co-pilot) and 12 cabin attendants, seven of whom were based in Washington and five in Beijing. Three of the Washington-based attendants and all of the Beijing-based attendants were bilingual in English and Mandarin.

After the door had been closed at Dulles and while the plane was backing from the gate, the cabin crew supervisor welcomed the passengers and then directed their attention to the short safety video which would appear on the monitors facing the seats. The video, with its accompanying audio playing over the cabin sound system, took about three minutes before concluding with a reassuring "Welcome aboard!".

All that could be done would be done to make the long trip a pleasant one for Flight 1230's passengers.

1.1 The ideal (shortest) flight between Washington, DC and Beijing tracks the Great Circle, the imaginary line on the globe between the two locations. For any particular flight the line will deviate due to weather and other conditions.

Flying nonstop from Washington-Dulles to Beijing is not an experience shared by many, and it is an everyday experience only for a very few airplane crews. Even so, Flight 1230 will provide a center point throughout the next six chapters for drawing our attention to several fundamentals of life which, among other factors, are pulling global populations together. Numbers, measurements, calendars, clocks, temperatures and climate are experienced by all of us. Usually taken for granted in a busy "just-the-way-it-is" world, they each deserve a day in the sun. Our climate, of course, is today's big challenge. To meet our new challenge, we should be asking what we might have learned from the earlier ones.

Departure

Cross-border trade has been a primary propellant of global unification. While Flight 1230 was aloft, on its way to Beijing, many cargo ships were plowing both ways across the Pacific, loaded with end products, parts for other end products and raw commodities. The production, transportation and distribution of goods, along with the many services that support those endeavors, were once viewed only through local lenses, but in the 21st century they have largely become multi-national in scope. Other propellants of global unification include conquest, colonization, military postings, diplomatic relations, leisure travel, education, missionary work, healthcare, and disaster relief. Cross-border ties, for whatever cause, spawn cross-border exchanges between people with different heritages and sometimes with conflicting values. The common thread is person-to-person interchange.

Inevitably, the countless person-to-person interchanges that have occurred over the course of human history, and are continuing to accumulate at an accelerating pace, generate changes within populations. Over time, those changes show themselves in everyday lives, sometimes only in very subtle ways. In the final analysis, all who expose themselves to this process of change, including those who never travel internationally, are both importers and exporters of customs and values. It is this ongoing process of import and export, a slow global percolation, that gives "just-the-way-it-is" an ever-changing meaning.

The video and audio systems enjoyed by Flight 1230's passengers during their long flight to Beijing embodied centuries of accumulated scientific ingenuity. And yet, they were perhaps among the least complex of the almost countless systems that are essential in a modern airplane. The fuselage, wings and air foils, the jet engines, the instruments in the cockpit and numerous other designs and devices from nose to tail all had much to do with safely lifting the aircraft off the ground, keeping it aloft, navigating and tracking its flight plan, and landing it safely at destination.

As fascinating as it is, air travel like that experienced by the passengers on Flight 1230 is only one piece of the much grander universe of achievements made by the many pioneering scientists who, over a broad span of time and around the globe, have put their marks on everyday life. Relatively few of these pioneers have received lasting recognition, but in the broad sweep of scientific advancement each individual contribution, in its own way, has been a steppingstone towards a better lifestyle for all.

Science is an international enterprise. Since its formation in the last half of the 19th century, the Bureau of Weights and Measures (BIPM) in suburban Paris has been in many respects at the hub of this enterprise, working in conjunction with national laboratories around the globe. (See the Appendix for a description of the BIPM and, more briefly, the national laboratories in the United States and a select few of other countries,) As succeeding chapters will indicate, neither the BIPM nor the national laboratories were in existence in centuries past, when many of the things in everyday life which we too easily accept as "just-the-way-it-is" were being puzzled out by our ancestors. They now play a critical role in scientific investigations. Although the climate per se is not a target of investigation at the BIPM or associated national laboratories, those institutions give direct support to scientists worldwide who study the climate by providing uniform standards of scientific assessment.

Chapter Two
NUMBERS

As previously mentioned, Flight 1230 departed Washington-Dulles on February 16 at 2:15 PM., providing nonstop service to Beijing. Its flight time was 14 hours, 15 minutes, putting it on the ground in Beijing on February 17 at 5:30 PM, local time. Total distance traveled was 6,915 miles.

The plane carried 233 passengers, including 112 Americans, 76 Chinese, 22 Canadians, 14 Latin Americans and 9 Europeans. In addition, on board were two cockpit crews (each a pilot plus co-pilot) and 12 cabin attendants, seven of whom were based in Washington and five in Beijing. Three of the Washington-based attendants and all of the Beijing-based attendants were bilingual in English and Mandarin.

Numbers Have Many Purposes

As the highlights above suggest, our numbers are multi-purposed. Pick any number from 1 through 9:

We all *think* about your number to be the same quantity. This is fundamental! Regardless of how any number is expressed, orally or in writing, using letters or a symbol, we all understand the same quantity.

We all *write* about your number, using a symbol. *1, 5,* and *10* are symbols. Our teachers wrote these symbols on the board. If we had grown up in China, our teachers could have instead put 一, 五, and 十 on the board. Same lesson, different symbols. In a completely unified world, the symbols would not be different

We all *write* about your number, using a word. *One, five,* and *ten* are words. In the English-speaking world, the teachers wrote these words on the board. In China, the teachers instead wrote 一, 五, and 十 (note, in Chinese the number symbols are also words). Same lesson, different words. Again, in a completely unified world, the words would not be different.

We all *talk* about your number, using an utterance. In English, we phonetically pronounce *one, five* and *ten*. In China, these numbers would be spoken as *i* (ee), *wú* (uwuh), and *shí* (schee) (parentheses show approximate phonetic pronunciations). Same meanings, different languages. And once again, in a completely unified world the language would not be different.

We all *calculate* with your number, using addition, subtraction, multiplication and division. This ties directly to our thinking. Same quantity, same calculation. This ability to calculate lies at the heart of much of the progress that humankind has brought to itself, and unfortunately also much of the violence it has inflicted upon itself.

Our shared thinking about numbers plus our shared ability to calculate with them are, in combination, a unifying phenomenon around the globe. Different populations still use different symbols and words, but even those differences in communication are diminishing in the rapidly increasing flow of international connections.

Numbers

Numbers have always had a dual existence. They are tools in everyday life but, in their own right, they also are a subject for intellectual study. As others have pointed out, we have numbers in life while, as any committed mathematician would assert, there is also a life in numbers. Humankind's long struggle with numbers vividly demonstrates that life and numbers have long been close companions.

Backstory

We count things with numbers, we record things with numbers, and we calculate things with numbers. The benefits of these three functions come to us, of course, because we also communicate with numbers. Although seldom appreciated as much as it should be, this shared knowledge of numbers is a really big deal for all of us. It did not come easily.

If we pause our busy lives long enough to learn a bit about humankind's starting points in the realm of numbers, we can be fascinated by the long, twisting paths that many prior civilizations around the globe have had to navigate, beginning about 30,000 years ago. The Sumatrans and Babylonians in Mesopotamia, the Phoenicians on the eastern coast of the Mediterranean, the Chinese, the Egyptians, the Greeks and Romans, various cultures in India, the Incas and Mayans in South America, and other ancient cultures all, in their own ways, had a part. Like those diverse and largely insular populations, the story about the development of numbers is a fragmented tale. And it is a story made even more difficult to recount by the relative scarcity of archaeological sources.

Boiled down, the history of numbers is that of many different systems, each having its own inception, slowly getting sorted out and ultimately converging into the single system used worldwide today. The most useful elements have been kept and the less useful, if preserved at all, have been sent into the archives. Before turning to the specific features that have converged into our modern number system, the long path of their discovery can best be highlighted by considering the various tools that different cultures have used to count, record and calculate.

As infants and toddlers, we used the most natural of counting tools – our fingers – to begin our learning: one finger, two fingers, three fingers, and then all the way up to five, eventually ten. This reflects how, thousands of years ago, shepherds kept track of their small flocks and farmers kept track of their small crops. Then as commercial trading, military logistics and other human pursuits came to require ever larger counts, first resort in many cultures was made to additional parts of the human body. For example, in what is now Paraguay the ancients, in their own language, used the utterance "finished both hands" for ten and "finished foot" for fifteen.[1] In some cultures, when the intended meaning was "many" the speaker would simply point to his hair.[2] Over the centuries, however, as the scope and pace of commerce continued to expand, the human body could no longer keep up with the growing demand for good counting, reliable records, and increasingly complex calculations.

In Europe, archaeologists have discovered animal bones bearing regularly spaced notches, dating from 35,000 to 20,000 BCE. In some cases the notches appear in clusters of five, like the fingers on a hand. This is early evidence of both an ability to count and a need to record whatever it was that ancient inhabitants were wanting to track. No tangible evidence of calculation appears with the notches, but something was surely being calculated in the heads that carved them.[3]

Numbers

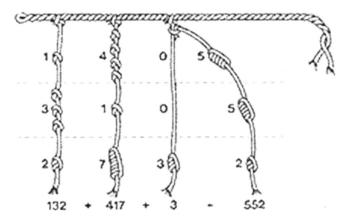

2.1 Knotted strings, Inca version (Arabic numbers have been added to show how the Incas were thinking)

Moving forward to the 16th century and crossing the Atlantic, Spanish conquistadores landing in South America found that the Incas had been keeping records not with notches but with a fairly complex system of knotted strings. The knots (*quipas*), by their relative locations on an arrangement of strings, were records that served specific purposes: commercial, administrative, liturgical and other. Even before the Incas used *quipas*, different versions of knotted strings had been used in Persia, Palestine and the Arab world. Again, the knots, or *quipas*, are tangible evidence of both counting and recording in ancient cultures. Calculation was no doubt occurring, but the knots themselves did not show it.[4]

Counting boards were yet another tool commonly deployed. Although the form varied, in essence the board was two or more rows of cells delineated on a table or other flat surface. Working from right to left across each row, markers were placed in the farthest right cell to indicate the number of units, other markers in the cell to the left to indicate units of ten, then others in cells continuing

2.2 Counting board

to the left to indicate units of 100, 1000, and so forth. The markers sometimes were nothing more than strokes drawn in a board covered with sand, and at other times they were pebbles or, especially in China, short rods of ivory or bamboo. Skilled mathematicians could rapidly perform calculations by shuffling the markers among the cells, doing the calculations in their heads as they moved across the array. When a calculation led to an empty cell that cell would be left empty with no markers, the equivalent of a place that would be a zero in our modern number system.[5] As will be seen again in the discussion that follows, the ancients were slow to conceive of zero as a place holder in a multi-digit number, much less as a number in its own right.

Numbers

The abacus is a direct descendant of the counting board. The classic version came from China in about the 14th century.[6] In the place of cells on a flat surface, a typical abacus consists of rods mounted side-by-side in a portable frame. Beads are affixed to each rod in a way that allows them to slide up and down to indicate a number, with units of one indicated on the first rod farthest to the right, units of ten on the rod to the immediate left, units of 100 on the next rod and so forth, right-to-left across the frame. Commonly, each rod has a bottom portion with five beads which can be located, some or all, to indicate any number up to five (in fingers, one hand) plus a shorter top portion with two beads, sometimes only one, which can be located to indicate the number five (one hand) as a proxy for a five in the bottom portion. Like the counting board, the abacus calls for practiced human skill, not to mention rigid mental attention, to efficiently perform complex calculations by sliding the beads through the correct sequence of arithmetic steps. The brain of the human operator is the calculator. The abacus is only the recorder of the results, step-by-step, in the calculation.

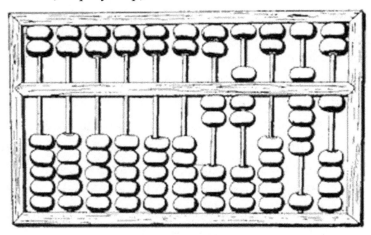

2.3 Abacus, Chinese version. Beads above the horizontal bar have a value of 5. The beads on the five rods to the right read 27091.

Over the centuries both the counting board and the abacus found wide usage in many cultures, having growing importance in the hands of merchants, government officials and others, not only in Asia but in the Arab world and Europe as well. Indeed, until the end of the 18th century the clerks of the British Treasury used these tools, called "exchequers" to reflect the arrangement of the rods and beads, to calculate taxes and for other purposes. (In Britain, the Minister of Finance is still called the "Chancellor of the Exchequer.")[7] The French Revolution, however, put the kibosh on the use of the abacus by setting the legislative stage to enact a ban on its usage in schools and government offices. This was another manifestation of the revolutionists' determination to reject the past and move France into the future.[8]

As mentioned, neither the counting board nor the abacus was a calculator. The same had been true for marked bones and knotted strings. The task of calculating was left to the brain of the individual in whose hands these tools were put to use. In this respect, there is an intriguing contrast between eastern and western cultures.

In China, those who had acquired the skill to do calculations on a counting board or with an abacus were highly-respected and often given important positions. Trust was easily given to their work, not only in commerce and governance but also in astronomy, astrology and other disciplines, as they were.[9] One example from ancient China adds a bit of color to this observation. Mathematicians were charged with ensuring the best line of succession for the ruling dynasty. This entailed programming the emperor's harem to take maximum advantage of astrological readings. It was worked out that every 15th night the empress would stay with the emperor, the next night it would be three senior consorts as a group, the next night nine spouses as a group, and then for subsequent nights 27 concubines in rotation, nine each night, until finally 81 female slaves would be called upon, in groups of nine. Then the empress would return to launch the rotation once again. As the Chinese mathematicians had it figured, all but the emperor were to have nights of rest.[10]

In Europe, a different attitude prevailed. As in China, few put themselves through the rigors of learning how to calculate. The training could be demanding — learning the basic operations of addition, subtraction, multiplication and division, as well as committing to memory the full menu of multiplication tables. For many centuries in Europe, those who made this effort were thought to possess supernatural powers and to be performing magic on their counting boards or with their abacuses. Even though their acquired skills were essential in many spheres of everyday life, the lack of even rudimentary mathematical understanding across most of the citizenry engendered mistrust and suspicion. Montaigne, the renowned French essayist who was educated in the classics, had once served as Mayor of Bordeaux and had been a friend of French kings, wrote with no embarrassment in 1588 that he could not "cast account [calculate] either with penne or Counters."[11]

Montaigne, like most of his fellow Europeans, was not yet aware that a new world of numbers had already come on shore. In time, new types of numbers and the ability to "cast account" were to become commonplace across Europe and eventually around the globe.

Arabic System

Our modern numbers comprise the so-called "Arabic" system. More accurately, we use "Hindu-Arabic" numerals. Beginning in the 8th century, as the Islamic religion was stretching westward across North Africa and eastward into Asia, the Arabs gradually adopted the number system they found in use by mathematicians in

India. Indeed, the Arabs historically called their digits, now ours as well, the "Hindu" numerals. They brought the essence of that Indian system into Europe near the start of the 10th century, when they had expanded Arabic rule into Spain.[12]

Europe as a whole in the Middle Ages was slow to accept the Arabic system. Under the dominance of the Catholic church, schools and universities dogmatically adhered to Greek and Roman learning, including the mathematics of those two ancestral cultures. Both of those cultures had used alphabetic letters, not distinctive symbols, to represent numbers (consider Roman numerals). This led to assigning mythological significance to numbers and that practice clouded the ability to think about numbers in an objective way. It had become common, therefore, for the ancients to see certain numbers as equivalents of the names of their gods and mythological figures. Thus, when the Middle Ages began numbers had become something spiritual and mystical.

This cloud lingered well into the Middle Ages. There was a general intolerance of secularism. Indeed, during the Inquisition of the 13th century those who used Arabic methods of calculation were often sent to the stake as heretics. Additionally, merchants and traders who had become accustomed to using Greek and Roman alphabetic numerals saw no reason to bother themselves with anything different.

Various forces, however, were cracking open European minds. After the Arabic system had been introduced into Spain, scholars, astronomers and others in a Europe that was only beginning to awaken from the darkness of the Middle Ages were finding a need in their respective pursuits for a more workable system of numbers. Those who were returning home from the Crusades in the 11th through the 13th centuries brought with them knowledge of written numerals and methods of calculation they had acquired from the Arabs. Meanwhile, translations into Latin from Arabic texts of Greek and Roman writings were finding their way into European universities, often carrying Arabic numerals within them.

The invention of the printing press in the 15th century accelerated the acceptance of Arabic numbers, but it was not until the 18th century, about seven hundred years after its first appearance in Spain, that the Arabic system was widely accepted across Europe. Likewise, that system was working its way into China and elsewhere in the Far East, eventually taking the place in many situations for indigenous systems.[13] Now, in the 21st century, the Arabic system has not only survived but has earned great credit as an enabler of many advances which have come down to us from the Renaissance, the Enlightenment, the Industrial Revolution and modern science. Most notably, the landing on the Moon is a spectacular example, but only one, of the empowerment humankind has found in its numbers.

The Arabic system has become so commonplace, so taken for granted, that thought is seldom given to the inherent features that have led to its essentially universal usage. Those features deserve a bit of sunlight.

Abstractions

Arabic numbers are abstractions. This is a mathematician's way of saying they can be used to count whatever it is we want to count. *Three* minutes (time - an intangible) are not the same thing as *three* sheep (animals - tangible), but the number *three* in both cases has the same meaning. The difference is just between time and animals; quantity is constant.

This observation, itself an abstraction, may seem perfectly obvious, but we should remind ourselves that the toddler who is taught to count "one finger," "two fingers," and so forth is likely for some time in early life to pair "one" with her index finger, "two" with her adjoining middle finger and so forth, attaching the number to the specific body part. At some point, however, that toddler comes to understand that "one" and "two" can be attached to anything, tangible or intangible. The Incas used different colors to make records with their knotted strings, depending upon what they were counting. And there are small populations around the globe that still use different number systems to count different things. For example, the Nivkh people, about 4,600 in total, on the mainland in eastern Russia and nearby Sakhalin Island, use 24 different sets of numbers in their native tongue. Long things like trees and pencils are counted with one set, flat things like leaves and textiles are counted with another, and so on.[14]

Symbols

The number symbols 1, 2, 3 ... through 9, now almost universally recognized around the globe, have evolved out of the somewhat different symbols the Arabs first encountered in the 8th century. They took their present form in the 15th century when the printing press was invented in Europe. Each symbol is an abstraction, not intuitively associated with anything that might be counted. And each is only a convention that has come to be uniquely associated with a quantity, nothing more. Together, our number symbols constitute a language of numbers, one that is separate from but usable alongside the many diverse languages of letters and words found around the globe.

Another set of Arabic symbols for our numbers coexists in parts of North Africa and eastward across Asia to Indonesia and Malaysia:[15]

Numbers

West Arabic Symbols   1 2 3 4 5 6 7 8 9

East Arabic Symbols
(alternatives 4 and 5)   ١ ٢ ٣ ٤ ٥ ٦ ٧ ٨ ٩
                                             ٤ ٥

Even in those regions where the East Arabic symbols are commonly used, there is a familiarity with West Arabic symbols, our symbols, especially among those who have ongoing contacts with the West.

As an interesting side note, we should ask in which direction, left to right or right to left, the East Arabic symbols are written and read when they appear within Arabic text. That text goes right-to-left, opposite the flow in the West. For example, we might write in English: "The University has 6,235 students." In Arabic text that would be written (staying in English, with apologies to our Arab friends): ".students 6,235 has University The". Two observations: first, within Arabic text the number symbols, unlike the text itself, are not reversed. As in the West, units are on the right, tens next to the left, and so forth. Second, a reader of Arabic text who encounters a multi-symbol number must jump across the array of symbols to capture the magnitude of the number, reading from left to right, before proceeding to read the text right-to-left. Why so?

Early Arabic scribes wrote left to right in successive lines on scrolls of papyrus stretched across their laps. They would then turn the scrolls clockwise a quarter turn to read top to bottom, starting at the right column and proceeding line by line, now positioned column by column, top to bottom, towards the left. This right-to-left order has endured, although since the advent of paper and pagination Arabic text has been written horizontally, but still right to left. When the Arabs later encountered in India, and then eventually adopted, the Hindu number system, the left to right direction in which the Hindu symbols appeared was incorporated into the right to left direction of Arabic text. The question is obvious: why did the Arabs not, from the outset, reverse the order of Hindu symbols so that they could be read in the same direction as Arabic text? Too late now.[16]

Evolving from the same Hindu source, both the East Arabic symbols and the West Arabic counterparts took different paths in the Islamic world. Two caliphates had emerged, an eastern one based in Baghdad and a western one in Cordoba. As it happened, those within the Baghdad caliphate in the East adhered more closely to the Hindu symbols while those in the Cordoba caliphate more freely made modifications consistent with the cursive styles of script prevailing in Africa and Spain. Result: same numbers, different symbols.[17]

In China, most people now primarily use the Arabic system, which did not find firm footing there until the beginning of the 20th century. China, however, has also retained its traditional numerals. They are used in finance, mainly for writing amounts on checks and promissory notes, and for some ceremonial purposes. Indeed, China has five sets of traditional symbols. The most commonly used is called "basic":

1	2	3	4	5	6	7	8	9
一	二	三	四	五	六	七	八	九

The other four sets include an official one to be used in financial documents, such as checks and promissory notes, to prevent fraudulent alterations plus three more with their own historical purposes. It is not unusual to find symbols taken from two or more of the five sets within the same document.[18]

Like many others that remain from ancient cultures, the Chinese sets include, as companions for the actual numbers 1 through 9, additional symbols to indicate orders of magnitude. These were needed to write numbers 10 or larger. The rank of each symbol in a multi-symbol string had to be shown by inserting a companion symbol showing its rank as a multiple of 10, or a multiple of 100, or of 1,000, and so forth. As discussed later, that need for a companion symbol of magnitude fell away with the advent in the Hindu system of positioning a symbol for zero where needed to indicate order of magnitude.

Numbers Are Language

As mentioned earlier, our number symbols together constitute a language of their own, separate from but usable with diverse languages of words. Symbols are used for calculations, as well as communications. As the history of numbers demonstrates, diverse populations around the globe, as their various cultures have become ever more interlaced, have recognized that they are well served by conforming to the number symbols used by others.

That recognition of shared benefit has been much slower to come in the many different languages of words. The notion of changing something so deeply implanted in a culture is for many a hurdle much too high. This means, among many things, that the different written and spoken words which diverse populations have attached to numbers remain firmly entrenched, even though the numbers themselves, by quantity, are the same. In written and spoken words, we cling to our own customary ways. Our grip on what has become familiar loosens ever so slowly.

Base 10

Our numbers use the base 10 (the decimal system). We count up to 9, but then to count further we need two digits, 1 and 0. And then after we reach 99, we need three digits, and so forth. The base 10 can be attributed to counting with our fingers, a practice that began millennia ago with our earliest ancestors and still gets much usage today.

But here we should pause to recognize that base 10 is not the only way to count. Looking backward in time, the Sumerians and then the Babylonians, beginning as early as 2500 BCE, counted things using the base 60 (the sexagesimal system)[19] and we continue to use this when dealing with spheres, such as the globe, and with circles, such as the clock face. Applying the base 60, the number 258 would mean (60 x 60 x 2) plus (60 x 5) plus (1 x 8). Converting to the base 10, that would be 7200 plus 300 plus 8, or 7508.

One complete trip around the globe, either east-west or north-south, traverses 360 degrees, while each degree of circumference includes 60 minutes of arc and each of those minutes includes 60 seconds of arc. Hence, the pilots on Flight 1230 were aware that their departure point at Washington-Dulles was latitude 38 degrees, 54 minutes north, at longitude 77 degrees, 2 minutes west. Flying the Great Circle, they navigated to Beijing for a landing at latitude 39 degrees, 55 minutes north, with longitude 116 degrees, 23 minutes east. All of these global coordinates carry the same assumption, which is simply that the pilots using them know the globe to be 360 degrees in its full circumference. Were it not for the influence of the Babylonians, perhaps the pilots could have more easily navigated around a globe divided into 100 degrees (using the base 10), not 360. As for the passengers on Flight 1230, what actually mattered was that numerical conventions existed and their pilots were in synch with them.

Similarly, one revolution around the face of a clock covers 60 minutes. Each day is divided into 24 hours (like base 60, the number

24 can be equally divided by the number 6), each hour into 60 minutes, and each minute into 60 seconds. It is a curiosity, however, that when seconds of time are divided further, our timekeeping devices, most typically our stopwatches, convert to a base 10, using the decimal system, to yield tenths, hundredths, and thousandths of a second. Research has not uncovered an explanation. Presumably, it lies in the history of the design of ever more precise timekeeping devices. But again, why not have a clock face divided into 100 minutes rather than 60? The globe of 360 degrees and the clock face of 12 hours (some show 24 hours), each hour with 60 minutes, are conventions now deeply embedded in our thinking.

For many of us, everyday life also involves the base 12 (the duodecimal system). When asked for our height, we answer in feet plus inches, it being understood that each foot includes 12 inches. When shopping, we look for a dozen (12) eggs, then maybe a dozen (12) red roses, or better yet a dozen (12) chocolate donuts! In Britain, until the units of currency were shifted to the decimal system in 1971, there were 12 pence (duodecimal) in a shilling but then 20 shillings in a pound sterling. Base 20 (the vigesimal system) is thought to be a vestige of the Mayan civilization. After reaching 10 on two hands of fingers, they continued with their toes.[20]

"0" Takes Its Place

Other Arabic numbers signify magnitude by their placement. When we write "1," we mean just one, but when we write "10" we mean ten, or "100" we mean one hundred. Two concepts are at work here.

The first is that we position the "1", which could just as well be any single-digit symbol up to "9", in the place that shows its order of magnitude: for example, 10 or 100. We use "0" as a place holder.

We could just as well have used any other single-digit to put "1" in its proper place but that would have produced a different number. We make good use of "0".

The second concept is that by using "0" rather than another number in our example we have communicated that there are no units in the number "10" and neither units nor tens in the number "100". That is an equally good use of "0". And this is a happy marriage. Zero plays two functions. It positions other numbers to the left into their proper place and, by doing that, it signals the orders of magnitude that have no value of their own. This marriage occurred in India before the Hindu-Arabic system was adopted by the Arabs and taken to Spain.[21]

Both Babylonian and early Chinese mathematics are thought to have used a place value system. The Roman numerals we still see on cornerstones and clock faces use place value, but like the ancient Babylonian and Chinese systems they do not use zero. In Roman numerals (alphabetic), a smaller number placed to the left of a larger one signifies a reduction of the bigger number to its right. Thus, CM means 900 (100 less than 1,000). But if the smaller number sits to the right of the larger, an increase is signified. MC means 1,100. The utility of "0" as a place holder, in comparison with the Roman system, can be shown in a simple addition:

Roman	Hindu-Arabic
CMIX	909
+CCIX	+209
MCXVIII	1,118

Deploying "0" as a place holder did not automatically cause it to be recognized as a number in its own right, alongside "1" through

"9". History stood in the way. In ancient Greece, Aristotle followed established tradition by defining "number" to mean a "heap", an accumulation.[22] It was then, and it still is, impossible to have a heap of nothing, so ipso facto "0" was not a number. When a counting board or an abacus was used to make a calculation, a cell or rod that showed no value was simply empty, a nullity, even though it did play its role in showing place value for number symbols in adjoining cells or rods.

The dilemma came when a number on a counting board or abacus was to be recorded on papyrus or paper. What would show the empty place, the nullity, so that the proper places would still be shown for the other symbols across the number in its entirety? Initially, place value was recorded in writings simply by leaving an empty space, but this often produced confusion or error. Then the empty place was recorded by putting a dot in its spot (this is still a practice when the East Arabic style of Arabic symbols is used).

Sometime during the 7th or 8th century, in all likelihood, mathematicians in India came to see that "0" should be given its own spot in their number system.[23] A nullity was indeed a number itself. While functioning not only to put other number symbols in their rightful places, a zero also communicated a value that gave meaning to the whole array. This recognition of the dual role of "0" made it possible to move calculation from the counting board and the abacus onto papyrus and paper, thereby making arithmetic accessible to all.[24]

Maybe we should wish that the ancients had chosen a symbol other than "0" to show zero. In our modern world we occasionally encounter alphanumeric arrays, such as those on products which designate model and serial. We are forced to ask ourselves if the symbol is a zero ("0") or a capital letter ("O")? Is it still too late? In Europe and elsewhere zero is denoted with the symbol "Ø".

Decimal Point

Our numbers use the decimal point. This allows us to use fractions with as much accuracy as a measurement or calculation may require, but with the concept of rounding to avoid the bother of unnecessary accuracy. (The period used in the United States to represent a decimal point becomes a comma in Europe. A curiosity, but contextually not a problem.) Without the decimal, how would a passenger at Washington-Dulles, before boarding Flight 1230 to Beijing, pay $12.99 for the paperback novel in the concourse bookstore?

Digital Device Addiction

Aside from base 10, with divergences into base 60 and base 12, we also use a number base consisting of only 0 and 1 (the binary system). In the long time scale of the history of numbers, our everyday use of the binary system is but a speck in time, running only from the later decades of the 20th century. Many of us have already become addicted. Few of us, however, have any awareness of the addiction. The addiction plays out, sight unseen, inside our computers and other digital devices.

The binary system has been foisted upon us in the computer age by the millions if not billions of electronic switches (on transistors and the like) in the guts of our pocket computers, smartphones, laptops, desktops and other digital devices. Each of the switches offers only two positions: open (no connection – the binary 0) or closed (connection – the binary 1). Software inside our devices converts not only our base 10 numbers but also the alphabetic letters, punctuation marks, symbols and operations we activate on our keyboards and touch screens (the "input') into assigned machine code sequences of 0's and 1's.

After the coded input has found its way through the multitude of switches, following program instructions which themselves shall have been coded into the device using binary numbers, the result (the "output") is then re-converted by software back onto the device's screen or a printer using familiar base 10 numbers and alphabetic characters we can read. The binary switches and instructions inside the device stay behind the scene.

The history of the computer age is itself a fascinating story, but not one to be told here. Instead, the question here is how the many digital devices we now keep close at hand might be changing our everyday lives. More specifically, how might these exceedingly fast and increasingly ubiquitous devices be affecting our capabilities to calculate without their aid? Will the day come when "arithmetic" begins and ends on a keyboard or touch screen? In the future, why would we need to learn the mechanics of addition, subtraction, multiplication and division?

The Higher Realm

On Flight 1230, the passenger in seat 24C was a statistician who had completed a three-month study of worldwide polio immunization reports housed at the National Institutes of Health in Bethesda, Maryland. She was returning home to Beijing to resume her ongoing research activities for Chinese health authorities. The passenger in seat 15A, a U.S. citizen, was a professor of theoretical physics at the University of Virginia who was going to Beijing to deliver a paper on gamma ray dispersion patterns at an international scientific conference. Either of these passengers, one Chinese and the other American, could tell us about the numerous ways that advances in the higher realm of mathematics have steered forward our still expanding body of knowledge about the one natural world all of us on the planet are sharing. Indeed, there really is a life in numbers!

Contemporaneously with ongoing advances in theoretical mathematics, serious efforts are underway in computer science to develop a quantum computer that will do calculations and other digital processing at exponentially faster speeds than current devices. We should all be prepared for the advances still to come, while we count on those who bring us ever faster devices to include, as they have so far, an easy way to use them with our Arabic number system.

Different cultures around our planet took many centuries to reach a common system of numbers, the one we call the Arabic system. Various cultures are still using different words and different symbols to speak and write about the same numbers, but the quantities behind the words and symbols are the same. And most importantly, that uniformity means calculations are the same. It is not far-fetched to think that as time passes and cultures become ever more integrated, the words and symbols used by different cultures will also become more uniform. That increasing uniformity will work to everyone's advantage. And we can be encouraged a much more recent development. The binary system, embedded inside our computers and other digital devices, represents a big step forward in our ability to wed electric power to communication and computation. Digital devices, using only the two digits "0" and "1", have opened our eyes to virtually unlimited possibilities for the future.

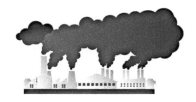

Chapter Three
MEASUREMENTS

Flight 1230 left Washington-Dulles, destination Beijing, at its scheduled departure hour of 2:15 PM. An essential part of the pre-flight preparations had been the calculation of the plane's gross weight at take-off. For this calculation, each of the 222 adult passengers was assumed to have a body weight of 195 pounds (88.4 kilograms) and each of the 11 child passengers 87 pounds (39.5 kilograms). The assumed total passenger weight of 44,247 pounds (20,088 kilograms) was added to the measured weights of the aircraft, the crew, the cargo, the fuel and other contents to assure that the fully loaded plane at take-off would not exceed its performance limit.

Covering the 6,915 miles (11,128.6 kilometers) to Beijing, Flight 1230 flew at various altitudes assigned by air traffic controllers. Its highest altitude along the route, in the Canadian control zone, was 39,000 feet (11,582.4 meters). After it had proceeded into the Chinese zone, the plane was stepped down gradually by traffic controllers for its landing at Beijing Capital International Airport, elevation 35 meters (115 feet).

The passenger in 37C, a native of Houston, was a post-grad PhD going to Beijing for his one-year appointment to teach American History at Peking University. When dinner service began, he purchased a half-bottle (375 milliliters) of chardonnay to accompany his choice of chicken parmesan. In the Economy section his seat pitch (distance between rows) was 31 inches (78 centimeters), hardly sufficient for his frame of 6 feet 2 inches (188 centimeters).

The passenger in 7A was a designer from Paris who would present her newest line at a showing in Beijing's fashion district. Along with a Caesar salad and prime rib, she enjoyed two pourings from a bottle (750 milliliters) of a favorite Bordeaux she had spotted on the dinner menu. In the Business section her seat pitch was 76 inches (193 centimeters), more than sufficient to accommodate her moderate height of 163 centimeters (5 feet 4 inches). After two exhausting showings in New York and Washington, she appreciated her restful overnight sleep during the long flight.

After landing in Beijing, the pilots of Flight 1230 determined that the aircraft's fuel tanks still contained about 18,750 pounds (8,503.4 kilograms) in reserve. One quick calculation told them this was equivalent to 2,757 U.S. gallons (10,436 liters). This fuel reserve was entered into the flight log, both to record the plane's performance on the just completed trip from Washington-Dulles and to inform the replacement crew that would be responsible for refueling in Beijing.

Units Are Conventions

Lengths, weights and volumes are the most commonplace measurements in everyday life. Like years, months, days, hours, minutes and seconds, these measurements of quantities, expressed with numbers, are essential connections among all of us, not only to coordinate our lives, living together as we do in a single global community, but also to propel us and future generations toward even better lives. We could not exist as we do without them.

As with calendars and clocks which track date and time, measurements of lengths, weights and volumes operate as widely accepted conventions about the sizes of the units in which they are expressed. The only magic in units of measurement is the shared understanding that is given to them across the planet's many populations. If alternative units expressing different sizes or using different names were instead in common usage, those alternatives would be equally good connecting tissues for society.

The story behind the evolution of the conventions of measurement now in widespread use is, in a very real sense, a reflection of humankind's ongoing effort to improve its existence. It is an evolution, however, that clearly has not yet fully unfolded. Before plowing into the historical roots of measuring and then tracing that evolution into modern times, it will be useful to put in front of us a snapshot of the measuring units and the measuring languages now in widespread use in the 21st century.

Two Measuring Systems

As indicated in the above account of Flight 1230, two systems of measuring units are in general use. First, the metric system is predominant both in everyday life in most regions of the world and throughout all of science around the globe. Second, the U.S. system (often called the "customary" system) remains predominant in everyday life in the United States. The story to come will explain how this duality came to be.

In the metric system, the basic unit of length is the meter (in the U.S. system, 39.3701 inches), the basic unit of weight is the kilogram (1,000 grams, equivalent to 2.2046 pounds), and the basic unit of volume is the liter (0.2641 liquid gallons, or in Britain 0.2100 imperial gallons). Metric units can be readily subdivided into smaller units or combined into larger ones simply by moving the decimal point and changing the prefix. A meter (1.000 meter) when reduced by a factor of one hundred becomes a centimeter (0.01 meter), or when reduced by a factor of one thousand a millimeter (0.001 meter). A gram (1.000 gram) when increased by a factor of one thousand becomes a kilogram (1,000 grams, or 1.000 kilogram). A liter (1.000 liter) when reduced by a factor of one thousand becomes a milliliter (0.001 liter). For science, especially, these easy decimal point conversions are both a major convenience and a safeguard against miscommunication. The benefits appear in everyday life, even in the United States, when a doctor writes a prescription, when a pharmacist fills that prescription, and in many other situations.

> "Meter" is the American spelling, while "metre" is preferred everywhere else. But there is an important exception: when referring to any device designed to measure time, distance, speed, or intensity or to regulate current, "meter" is the preferred spelling everywhere.

In the U.S. system of measurement, the basic unit of length is the foot (in the metric system, 0.3048 meter), the basic unit of weight is the pound (0.4536 kilograms), and the basic unit of volume is the gallon (3.7854 liters). Each unit has its own subdivisions (a foot can be subdivided into 12 inches) and enlargements (a foot can be tripled to become a yard), each subdivision or enlargement having its own name. The numerical relationships between one subdivision or enlargement and another must be known to convert a single measurement (say, 2 feet, 4 inches) into another (28 inches, or 0.778 yard). Similar but different subdivisions, enlargements, and relationships exist for units of weight and volume. Neither the convenience nor the communication safeguard of the metric system is inherent in the U.S. system.

Making conversions between the two systems is not difficult, assuming both an awareness of the need to do so and the existence of basic mathematical skill. A human brain is fully capable of making those conversions, whether that means resort to a conversion tool on the internet, key punching on a pocket calculator, or working with pencil and paper. The risk in making a conversion lies mostly in the assumption of an awareness that a conversion needs to be made. There are tragic tales to tell about the hazard inherent in that assumption.

> "Weight" is the result of stuff being acted upon by gravity. "Mass" is simply the amount of stuff. When Neil Armstrong went to the moon, his weight decreased as he moved farther away from the Earth's gravity and nearer to the moon's weaker pull. His mass, however, did not change. Same stuff, less gravity.

One unfortunate tale involves the refueling of a Boeing 767 aircraft for its Montreal to Edmonton flight on July 22, 1983. The amount of fuel loaded in Montreal had been incorrectly calculated due to confusion between measurements in the British imperial system and the metric system (in Canada, the law in 1979 had changed sales of fuel from imperial gallons to metric liters). Less than half of the intended amount of fuel was in the tanks at take-off. Both engines ran out of fuel and they unexpectedly shut down when the plane was at altitude. That loss of power likewise caused a loss of hydraulic controls and nearly all instrumentation. Skilled pilots were able to glide the craft to an emergency landing at a decommissioned Canadian Air Force base, at the time a drag race track, in Gimli, Manitoba (the incident is sometimes called the "Gimli Glider"). Their skills and a large measure of good luck prevented any serious injury for passengers or crew, but the measuring error in Montreal certainly could have had far more tragic consequences.

A second tale, less dramatic but definitely illustrative, involves the September 1999 disappearance of a $125 million Mars orbiter which had been launched by NASA in 1998. The entire program was designed using metric units. Critical performance data, however, had been provided in U.S. units. The end result was that the orbiter entered the Mars atmosphere well beneath its design trajectory, presumably causing the craft to crash rather than orbit Earth's neighboring planet. That was an extremely expensive ooops!

BIPM Units

Two of the BIPM's base units of measurement are length and mass (the BIPM and scientists use "mass" when most of the rest of us would less precisely say "weight"). Volume, for which the BIPM standard is a cubic meter (equivalent to 1,000 liters, or 1,000,000 cubic centimeters), is one of the BIPM's derived units, involving as it does a multiple of three units of length.

Among the BIPM's seven base units, the only one which has not yet been expressed in metric units is its standard for mass, what most of less precisely call weight. As previously mentioned, that standard has long been a physical one in that it has been based upon a casting of alloy metal in the BIPM's custody outside Paris. In November 2018, the CGPM, the international body which oversees the BIPM, gave final approval to a redefinition of "kilogram", along with redefinitions of three other measurement standards. Replacing the physical cylinder of metal in the BIPM's safe outside Paris, the newly adopted standard for the kilogram is one which can be derived from what scientists call Planck's constant, a computed measurement of packets of energy at the quantum scale. Few among us need to know the details, but scientists who have committed themselves to the precision of measurement see this to be a very important advancement.

Backstory

The story of measurements, which now in the 21st century converges at the BIPM, has multiple starting points. In ancient times, when contacts between cultures were neither frequent nor essential, each population developed its own system of measurements and its own names of units. Scant attention, if any, was given to systems used elsewhere. Even within a given population, there was only crude standardization.

In ancient Greece the human foot (the *pous*) was a common standard for measuring length , but it was an inexact standard and, within a culture with numerous localities, there was no universal understanding of its size. Likewise, in ancient Rome the foot (the *pes*) was a standard of length, with the same lack of common understanding. In ancient China the measurement of length most closely approximate to a foot was the *chi*, the length of which was extended by successive dynasties from about 6 ½ inches during the Shang Dynasty (second millennium BCE) to its current dimension of about 13 inches in the traditional system still in everyday use there.

Aside from measurements of length, parts of the human body were also used as standards to measure volume (even today, a recipe may call for a pinch of salt). Quantities of grains and seeds were measures of weights. In Britain and Ireland, for years past and continuing today, human body weight is measured in "stones", one stone being 14 pounds or about 6 1/3 kilograms. These standards and others served their purposes within the localities where they were developed for the needs of the local population.

But needs changed as populations and trade grew. People became ever more interdependent. In Europe, in particular, the feudal structure of the Middle Ages took hold. Feudal lords sought ways to impose control over those who worked their land and those who defended their fiefdoms. One mode of control involved the adoption of a unifying system of measurement for all who were expected to be obedient to their master. In various ways, those systems served as everyday reminders of where obedience was to be given. The lord's arm might be one standard of length, 100 paces of the lord's step might be a longer unit of length, and so forth. Above the lord and his own fiefdom, a king, queen or other authority might declare a separate set of standards.

As can well be imagined, as trade increased between localities using different standards there was likewise an increase in the chaos and frequency of errors in measurements. Within a fiefdom, the lord was in control of measuring the crops produced and revenue derived from the land assigned to a particular peasant. When the crop was delivered for the lord's account to a merchant at the nearby market, that merchant might well have shorted the lord by using a different standard to weigh the grain or other produce. Worse, opportunities for abuse became more common, and abuses happened. In the more extreme cases those abuses could lead to truly harsh punishments, such as cutting off fingers or even execution. Even when error on a measurement was not intentional, it came from the confusion that was inevitable in the jumble of measuring standards and terms that had arisen from a multiplicity of local practices.

The French Revolution

Discontent with the feudal structure festered across Europe in the late Middle Ages. Most notably, the peasantry in France in the 17th and 18th centuries became increasingly resentful. The poverty, chaos, and individual abuses the peasants were enduring, against the backdrop of a Crown that was imposing heavy taxes to pay for its own regal lifestyle and military adventures, brought calls for action.

King Louis XVI must have been dozing on his throne. It was not until 1789 that he invited nobility, clergy and commoners to express their grievances. The complaints that came forth included the abuses arising from the prevailing diversity of weights and measures. That complaint and the insistence upon a uniform system of standards became one of the battle cries of the French Revolution. That social and political revolution, of course, was an upheaval with multiple causes. It led to King Louis' execution in 1793 and then to Napoleon's declaration in 1804 that he was the Emperor.

A fervor for conversion to the metric system seized the revolutionists, no doubt in part because it reflected their zeal to rid France of all they saw wrong in their nation's past. In the background of all the mayhem that befell France during the 1790s, advocates for metric conversion were maneuvering to make that happen. It was far from a fickle cause.

Scientists in France and elsewhere understood the benefits of a metric system. Their work was hampered when they needed to communicate with fellow scientists elsewhere who were using different standards and terminology. The Enlightenment had taken hold. Uniformity would be a big plus for the efforts then underway to find pathways to scientific discoveries and technical progress.

The demand for a uniform system of weights and measures was taken up by both the Royal Academy of Sciences and the succession of national legislative bodies that oversaw French lawmaking during the revolution. The Academy adopted the name "meter" for the basic unit of length and, connecting that name to the decimal system of numbering, it created subdivisions and multiples of that basic unit. With that, the Academy embarked upon determining the exact length of a meter by measuring the distance of a sector of Earth's meridian, from Dunkirk to Barcelona (10 degrees of latitude). The length of the meter would be one forty-millionth of Earth's full meridian. (Why one forty-millionth? No explanation has been found.) That basic unit of length was then subdivided into centimeters, and then using that shorter unit of length a cubic centimeter of water would become the gram, the basic unit of weight. The gram is the weight of a cubic centimeter of pure water at the temperature of melting ice (a nudge above 32 degrees Fahrenheit, zero Centigrade).

As events played out, including disruptions attributable to the political turmoil of the revolution, it took the Academy seven years, until 1799, to make its determination of the length of the meter and the weight of a gram. In the course of its work, a decision was made that a kilogram, rather than a single gram, would be a more practical standard by which to establish, preserve and promote the standard for weight.

Meanwhile, the French legislative body in 1795, then named the National Convention, announced that the new metric system would be mandatory and the basic units would be the "meter," the "gram," and the "liter" (cubic decimeter). In the same step, it announced that prefixes would be used for a comprehensive nomenclature: subdivisions of tenth, hundredth and thousandth, multiples of tens, hundreds and thousands. These declarations were made years before the actual sizes of the new units were to be finally determined, which would have to await completion of the Academy's work to determine the length of a meter. Apparently, those serving in the legislative body felt a compelling need to take action towards a uniform set of standards, in response to the political fervor for reform. (Later in 1795, the National Convention abolished the Royal Academy of Sciences and reconstituted it as the National Institute of Sciences and Arts. Years later the name would be changed again, making it the French Academy of Sciences.)

As completion of the Academy's work was approaching, France invited other nations to send delegates to Paris to participate in the preparation of the final report about conversion to the metric system. Spain, Denmark and other European nations accepted. The foreign delegates who participated outnumbered those from France. Hence, the report that went to the French legislature in 1799 was the product of what might be called an international gathering of scientists. The French legislature adopted the report, which had been more than seven years in the making.

Physical prototypes of the meter and the kilogram were made the official and mandatory standards for all of France. Made of platinum, they became known as the Meter of the Archives and the Kilogram of the Archives. (These first prototypes still exist in the National Archives in Paris, but their duties as standards of measurement have been supplanted.) The reformed nation of France was proud to show itself as a leader of science and technology. It took nearly 50 years, however, for the metric system to become firmly implanted across all of France. The new metric system was used side-by-side with the various pre-revolutionary units of measurement, thus leaving in place much of the confusion, and presumably many of the abuses, that had been one of the sparks which had ignited the French Revolution. It was not until 1840 that the use of non-metric units of length, weight and volume became illegal in France.

Just so, the metric system we know today is an important legacy of the French Revolution.

Metric System Followers

Napoleon's forces took the metric system along as they conquered much of Europe in the early years of the 19th century. Some places readily received it. Others had to be persuaded, in ways perhaps that only a conquering army can deploy. It can be assumed that the scientific community needed much less persuasion than the majority of citizens, accustomed as general populations had become to their traditional measurements.

By 1850, Belgium, Luxembourg, The Netherlands and Algeria (then a French colony) were on board. By 1900, the count included 33 more countries, including Mexico and most of the countries in Central and South America. But Britain, Canada and the United States were not on that list. Nor was China or any other Asian nation. The metric pot still needed more time to cook.

Britain Took Its Own Path

Like France, Britain in the early 19th century, with its robust industrial growth, saw the need for a standardization of measurements. As in France and elsewhere, the lack of standardization had become a source of chaos and, it can be assumed, at least occasional abuse. For example, England and Scotland, early in the 19th century, had been unified politically for more than 100 years, and English units of measurement from the time of that unification were supposed to be the standards. In everyday life in Scotland, however, the Scottish measures of pints and gallons, more than twice the size of English pints and gallons, had remained in common use, along with the English measures.

During the 17th and 18th centuries, several British scientists had advocated adoption of a system of measurement based upon decimals. They had a full appreciation of the benefits that would come from working with units that could be subdivided and enlarged simply by moving the decimal point and changing the prefix. More than that, they wanted a system that would make their work more easily understood by their fellow scientists in France, Germany, Italy and other countries. Their advocacy for a decimal system had good reasons behind it, but no change in British law, much less everyday British life, came to be.

Lawmakers in Britain tackled the need for standardization in 1824, but they did so not by following France into the metric system. Instead, Parliament defined units of measurement for Britain's own "imperial system." The new system gave a fixed meaning throughout the British Empire to units of length (some familiar in the United States, such as inch, foot, yard and mile, but also some not so familiar such as thou, chain and league), units of weight (familiar ones such as ounce, pound, and ton, plus not familiar such as drachm, stone and quarter), and units of volume (familiar such as ounce, pint, quart and gallon, plus the not familiar gill). In addition to these units of length, weight and volume, the 1824 legislation defined units of area (perch, rood and acre) and navigation (fathom, cable and nautical mile). Britain, the global leader in industry and trade throughout much of the 19th century, had decided to do things its own way.

Advocacy for conversion to metric continued throughout the 19th century. Legislative efforts in that direction from time to time came close to passage but never made it over the bridge. In 1875 a British delegation joined those from 19 other countries in Paris to negotiate the Meter Convention of 1875, which established the BIPM and the international bodies that oversee it. Britain, however, declined for two years to sign that convention. Eventually, in 1896, Parliament approved the use of metric units for all purposes, but it stopped short of making their usage compulsory. By that time nearly one-half of British exports were going to countries using the metric system. Nevertheless, everyday life in Britain went on in imperial units, not metric units. Unknowingly at the time, that 1896 approval of metric units would become essential in the last half of the century that followed.

Wars and depression, plaguing the first half of the 20th century, took attention away from the issue. It did not find focus again until about 1950. Soon thereafter, a report to the British government unanimously recommended a compulsory conversion to the metric system, as well as decimalization of the currency, both to occur within ten years. Initially, that recommendation was rejected by British industry, but in 1963 the government was informed that industry favored the metric system (some industrial sectors, however, wanted voluntary, not compulsory, conversion).

Beginning in 1965, the British Standards Institution (BSI), in its role as Britain's body for setting commercial standards, coordinated the efforts of various industrial sectors to bring the metric system into British commerce, while The Royal Society, an independent scientific academy, worked with professional societies, schools and the like to educate the public. Fixing 1975 to be the target for conversion to metric, the government created an advisory Metrication Board and announced that there would be no compensation for making the conversion. Costs would be borne where they fell.

The opposition did not fall silent, but throughout the 1970s several industrial sectors proceeded with metric conversion. To fix on only a few, by 1972 both the paint and the steel industries had completed the process, and they were followed in 1975 by both retail trade in fabrics and floor coverings as well as medical practice. In some cases, a sector made the conversion to metric but also chose to retain the imperial system.

Contemporaneously, the metric system was being taken into British classrooms. In Scotland, all examinations from 1973 onward used metric units. In England, each examination board set its own timetable, with science and mathematics using metric by 1972, geography in 1973, and home economics and various crafts by 1976. Metric was the principal system when the national curriculum was introduced in English schools in 1988.

While industry and education were making their efforts towards conversion, on a grander scale all of Britain had its eye on possible membership in the EU (called the European Economic Community until 1993, then renamed the European Union and now the EU). In 1971 the EU, then still having only its six founding member nations, issued a directive requiring its members to use the metric system as defined by the BIPM, with only limited exceptions. Britain became a member of the EU at the beginning of 1973, along with Denmark and Ireland. The EU directive gave the three new members five years to complete their conversions to metric. Differing viewpoints have been asserted about the impact that prospective and then actual EU membership might have had on the conversion efforts within Britain.

Britain fell short of meeting the EU's 1978 deadline. Momentum for conversion had declined. In 1980, the earlier EU directive was replaced, giving more time to comply in several important respects. A number of traditional units of measurement that had been proscribed by the earlier directive could remain in use until 1985. Beyond that, a number of imperial units, including basic ones of length, weight and volume, could continue to be used until the end of 1989. And even more forgiving, until Britain might determine otherwise its imperial units could continue to be exclusively used on road signs, for the sale of milk in returnable containers, for the measurement of draught beer and cider and for land registration. Denmark and Ireland received similar concessions.

The 1980 EU package gave the British relief from a confrontation with EU authorities. In the viewpoint of many British citizens, there was no realistic prospect of across-the-board conversion, no matter what any EU directive might require. From 1980 forward, there have been more EU directives allowing additional relief. (As now well known, in 2016 British voters approved a referendum for their nation to leave the EU. By 2020 the separation of Britain from the EU became fact.)

A turning point in the British conversion effort had come in 1977 when a major London retailer of floor coverings found a sales advantage by reverting to imperial units. When carpet was priced by square meter it appeared to consumers to be more expensive than if priced by square yard, even though in absolute terms the two prices were equal. Consumers did not understand that the square meter, a metric measure, was larger than a square yard, the imperial measure. Parliament, industry and educators may have chosen to convert the country to metric, but consumers were still anchored in the imperial units they had long been using.

Industry groups reacted by arguing even more vigorously for Parliament to fix a cut-off date, making conversion to metric mandatory for all. But in the end, the political will was not there. A general election was coming in 1979. An order to enact a cut-off date was in the hands of the Labour Government ahead of the election, but it was never submitted to Parliament for a vote.

In a very real sense, Britain is still a nation that uses two systems of measurement, both the imperial and the metric. The size of each imperial unit is defined in terms of its metric equivalent, but the use of imperial units is widespread in everyday life. Most British people use imperial units for distances, body heights and weights, and volumes. Scales which display both units are commonplace in retail trade. Many pre-packaged foods carry dual labeling, imperial beside metric. Clothing is sized in inches, but often with meters as supplementary information. Road signs must use imperial units and vehicles must be equipped to show speed in miles per hour. Gasoline (petrol), however, is sold now at the pump in liters. Previously priced by the gallon, gasoline unit pricing changed to the liter when the price rose above £1.999 per gallon, the design limit on many pumps. A liter is only about one-fourth of a gallon, and that difference meant that the pumps already in place would still work if gasoline were to be sold by the liter, not the gallon. Finally, but perhaps most importantly, at day's end draughts of beer and cider must be ordered in pints (the imperial pint, about 25 percent more than the United States pint). Some things are too British to be changed. On the other hand, metric units have become predominant in most industrial and commercial activities and by government and public transportation systems. The London underground uses metric, as does the Chunnel tunnel running between England and France.

Conversion to metric has been sluggish in Britain. And to complicate the matter even more, Britain's exit from the EU raises questions about its ongoing adherence to most but not all EU standards of measurement.

Early America

Colonial settlers brought with them the units of measurement they had used in England and other native lands. The thing to have in mind, however, is that until the first half of the 19th century, two centuries after the Mayflower, there were no uniform standards in those native lands, with the result that the different colonies, and for decades after the American Revolution the different states as well, had no uniform standards. The principal units were called the yard, the pound, the gallon and the bushel, but divergences in the sizes of even these units were common.

The need for uniformity was evident. The Articles of Confederation, ratified in 1781, and subsequently the Constitution, effective 1789, explicitly gave Congress the power to provide uniformity. Article 1, section 8, of the Constitution said when written and now still says: "The Congress shall have the Power ... To ... fix the Standard of Weights and Measures." But this is a constitutional power yet to be fully exercised, in spite of several urgings that Congress step up to the plate.

In his first message to Congress in January 1790, President Washington declared that the uniformity of weights and measures was a matter of great importance and stated his presumption that it would receive due attention. No action. Thomas Jefferson followed with a set of proposals, one to fix standards for units in the U.S. system and, in the alternative, to fix new units in the metric system. No action. A Senate committee report in 1792 recommended the adoption of Jefferson's alternative proposal. No action. Years rolled by. In 1819 a committee of the House of Representatives proposed adoption of standards conforming to those in common use. No action. A report supplied to Congress by John Quincy Adams in 1821 recommended standards based on units commonly in use. No action.

It can be argued that this Congressional restraint was based upon the ongoing confusion still being experienced in both France and Britain, where efforts to establish uniformity were struggling for broad acceptance. Why impose that confusion on a new nation, a young country which still had seemingly more immediate issues calling for attention?

But Congress was not entirely passive. In 1797, in its first legislative action on weights and measures, Congress ordered the collectors of customs at U.S. ports to apply standards in assessing import duties. The difficulty, however, was that there were no such nationwide standards to be applied until thirty-five years later, when in 1832 they were established by the Treasury Department. That was done in compliance with a resolution, not a law, passed by the Senate. Large discrepancies of measurement were then discovered among the ports. A degree of uniformity was established, but only for the collection of duties. The standards for measuring length (the yard) and weight (the pound) established by the Treasury Department in 1832 for the collectors of customs were in conformity with their British counterparts, both before and after Parliament in 1824 had fixed the British standards. The standard that Treasury established for measuring volume (the gallon), however, was different than its British counterpart. This came about because until 1824 there had been three commonly used standards for the gallon. The one that had taken

> The United States has a separate standard for the so-called dry gallon, a measure of about 269 cubic inches. This is the legacy of the third standard that pre-1824 Britain used, the corn gallon. As its name suggests, the dry gallon was once the measure for corn and other dry commodities. It has no current commercial use. In this book, references to "gallon" are about the liquid gallon (either the smaller U.S. measure or the larger British measure).

precedence in the American colonies was known as the wine gallon, or Queen Anne's gallon, a measure of 231 cubic inches. When Parliament fixed standards for Britain in 1824, it opted to make the imperial gallon, what had been known as the ale gallon, a measure of 282 cubic inches. (Yes, these British names of yesteryear do say something about lifestyle.) Consequently, in a twist of history, the gallon measure in the United States is only eighty-two percent of its counterpart in Britain.

In 1836, both chambers of Congress, acting by resolution, not new law, directed the Treasury to supply sets of standards, in the form of physical copies, not only to collectors of customs but also to each state, "to the end that a uniform standard of weights and measures" may be established throughout the United States." This was followed in 1838 when, as an amendment to an appropriation act for the U.S. Military Academy, Congress directed the Treasury to distribute standard weight balances, or scales, to each state.

Within the Treasury Department, the Office of Weights and Measures was established to carry out these Congressional instructions. (This function, originally within the Treasury Department, is now in the NIST, along with its many other responsibilities.) By 1850, practically all states had been supplied with the physical standards set by Treasury. As new states were admitted each was also presented with those standards. Congress itself, however, still had not exercised its constitutional power to fix nationwide standards. That left the matter in the hands of individual states. Apparently, the needs of commerce in a rapidly growing economy compelled the states to conform their measurements to the standards that had come from the Treasury Department. Economic growth continued, with some rough bumps along the way.

U.S. Metric Gate Opens

While the nation in its early decades had been busy becoming a nation, some who had an interest in doing so had been watching the conversion to metric in both France and an increasing number of other countries. As might be expected, the National Academy of Sciences took an interest and recommended action by Congress.

This time, Congress did listen and respond. The Metric Act of 1866, permitting (but not mandating) the use of the metric system of weights and measures, became law after the chairman of the House Committee with responsibility for weights and measures had explained its purpose. Quoting in part:

> *"The metric system is already used in some arts and trades in this country .... Its minute and exact divisions specially adapt it to the use of chemists, apothecaries, the finer operations of the artisan, and to all scientific objects. It has always been and is now used in the United States coast survey. Yet in some of the States, owing to the phraseology of their laws, it would be a direct violation of them to use it in the business transactions of the community. .... <u>The interests of trade among a people so quick as ours to receive and adopt a useful novelty, will soon acquaint practical men with its convenience. When this is attained—a period, it is hoped, not distant—a further act of Congress can fix a date for its exclusive adoption as a legal system.</u>"* [Emphasis added. It has been a "period" very distant, indeed! Now, more than 150 years later, the anticipated widespread usage has still not happened!]

Years before, the Treasury Department had received physical copies of the meter and kilogram embodying the standards of metric measurement that had been adopted in France and elsewhere. (See image 3.1 above.) In conjunction with declaring the metric system to be legal, Congress in 1866 directed that copies of these standards be delivered to each state. That task was completed by 1880.

The timing was fortunate. This action by Congress in 1866 assured the United States a seat at the table in Paris when, in 1870, delegates from twenty countries gathered, at the invitation of the French government, to consider the establishment of an international body to oversee the metric system. That initial conference, although itself short in duration, led to the signing of the Treaty of the Meter in 1875, with the United States being one of the seventeen initial signatories. Britain, Greece and The Netherlands tagged along later. The Treaty of the Meter spawned the formation of the BIPM.

The BIPM, outside Paris, as a first order of business attended to the making of prototypes of the meter and the kilogram, taking as much care as possible to assure accuracy and stability. The pair that agreed most precisely with the meter and the kilogram that had been deposited in 1799 in the French National Archives were declared to be the international prototypes. Copies of that chosen pair were distributed as national prototypes to the countries that had become parties to the Treaty of the Meter.

The United States received two pairs in 1890. In a White House ceremony befitting the significance soon to come, one pair was accepted by President Harrison and declared to be the national standards of the meter and kilogram. They were then housed for safekeeping with the Office of Weights and Measures. The second pair reached that agency a few months later, for deposit alongside the first pair, the national standards.

The name Thomas Corwin Mendenhall (1824 – 1924) deserves a special place in the hearts of those committed to the metric system. He was the Superintendent of Weights and Measures in 1893. Acting with the approval of the Secretary of the Treasury, Mendenhall published, as Bulletin 26 of the Coast and Geodetic Survey, his decision that the international meter and the international kilogram would be the fundamental standards of length and weight (Mendenhall more accurately called it mass) in the United States.

3.1 Thomas Corwin Mendenhall (1841 – 1924)

This decision was given more permanence by republication as an appendix to the 1893 annual report of the Coast and Geodetic Survey. That report also included tables for making conversions between the metric and U.S. systems.

Called the Mendenhall Order, this bureaucratic action set to the side the difficulties that had been encountered during the 19th century in keeping the standards of the U.S. system in line with those of Britain's imperial system. It also had the effect of sweeping into the United States the past and future efforts of the BIPM to establish and maintain internationally uniform standards. Even so, Americans continued to measure, write, speak and calculate about yards, feet and inches, not meters, about pounds and ounces, not kilograms, and about gallons, quarts and pints, not liters. There is scant awareness that the Mendenhall Order made the U.S. system at its foundation a stepchild of the metric system. But the underlying fact is that it did just that! As a direct consequence of the Mendenhall Order in 1893, the actual sizes of units in the U.S. system (for example, the yard, the pound and the gallon) have been derived from the sizes of units in the metric system (the meter, the kilogram, and the liter). For more than a hundred years the United States has been clinging to its customary units in name only.

Not too many years after the Mendenhall Order, Congress did respond to lobbying by the American scientific community and its supporters in industry. In 1901 it moved the Office of Weights and Measures, until then an agency of the Treasury Department, into a newly established National Bureau of Standards within the Commerce Department. That new agency then promptly adopted the Mendenhall Order, making the metric system a basic component of official standards in the United States, even if not a visible one in everyday life. With this administrative action the long dereliction by Congress of its authority under the Constitution "To … fix the Standard of Weights and Measures" became irrelevant. The National Bureau of Standards was renamed in 1988 to be the National Institute of Science and Technology (NIST), reflecting its greatly enlarged scope of responsibilities.

World War I, depression, and then World War II took public attention away from units of measurement in the United States, just as they did in Britain, the rest of Europe and most of the world. The early decades of the 20th century were highly productive ones for science, with the work of Einstein and many other luminaries who were puzzling out relativity, quantum mechanics, and other frontiers. But the long-term importance of the work of the world's community of scientists was beyond the field of vision of most mortals.

Technological advances were much closer. The early decades of the 20th century brought lasting advantages to everyday life with the rapid spread of electric power, telephones, radio, automobiles, aircraft, and other lifestyle improvements, all of which are now taken for granted. Americans, in particular, but much of the rest of the global population as well, were preoccupied with the technological improvements coming into their lives. At the same time, during the first half of the century they were often focused, in both their private lives and their public ones, upon the struggle to keep a troubled world from tearing itself apart. There was little time to be concerned about units of measurement.

Voices for conversion to the metric system began to receive attention again in the second half of the 20th century. In 1968, three years after Britain had begun its transition to the metric system, Congress authorized a three-year study by the Department of Commerce about the feasibility of conversion in the United States. An advisory panel that participated in that study recommended in 1971 that the United States follow the path already chosen in Britain, and other countries as well, by making a conversion to metric over a period of ten years.

Recall that Britain in the 1970s was in the throes of metric conversion, in part with its eyes upon EU membership. Likewise, the level of interest in the United States reached its peak in that decade, before interest in metric started its slide in the late years of the decade, back into the list of lower priorities.

While interest was still at its peak, in 1974 Congress passed and the President signed a comprehensive educational bill which included language encouraging schools to prepare students to use the metric system. That 1974 legislation was replaced in 1978 by a law authorizing financial support for the same purpose. In a more direct but only half-hearted response to the 1971 recommendation for a mandatory conversion to metric, Congress passed the Metric Conversion Act of 1975 with the stated purpose being "to designate the metric system of measurement as the preferred system of weights and measures for United States trade and commerce; ...." The shortcoming of the 1975 legislation, however, was its failure to make conversion mandatory. It set no schedule. The Act did establish the U.S. Metric Board to oversee conversion, but that body was buried seven years later after the Board complained to Congress that it had not been given a clear mandate.

Congress acted again in 1988, but with no greater force or impact. In that year it included in the Omnibus Trade and Competitiveness Act another declaration that the metric system was preferred and it required federal agencies to use that system "to the extent feasible" by the end of fiscal 1992. This legislation was followed by an executive order in 1991 that directed all federal agencies to "take all appropriate measures" (we should assume no pun was intended) to use the metric system in "business-related activities." That executive order, however, in its succeeding section qualified the directive:

"Metric usage shall not be required to the extent that such use is impractical or is likely to cause significant inefficiencies or loss of markets to United States firms."

The barn door allowing escape from both the underlying mandates of Congress and the directive of the President was too wide. Many agencies did little to convert to metric.

United States Using Two Roads

The nation is still very much traveling down two roads. Science and applications that follow it are going down the road of metric units. That road is encountered by consumers perhaps most commonly when they obtain a medical prescription. Medical practice uses metric units. In other aspects of everyday life, however, American consumers are staying on the more familiar road of U.S. units. Those units are used, for example, to measure body weights and heights, purchase food and clothing, and fill gas tanks. This being the case, retailers, manufacturers and other businesses, both domestic and foreign, that target United States consumers are typically continuing to label their products with U.S. units. At the same time, in an increasingly globalized economy these same businesses may well be using metric units to target other consumers elsewhere.

The United States is often said to be a nation that has not yet "gone metric." (In this respect, the United States is lumped only with Myanmar (Burma) and Liberia.) This assertion, however, is not entirely accurate. A slow conversion to metric has been underway in the United States for several decades. It differs only by its degree of sluggishness from that in Britain. The difference between the two nations can be attributed to the stronger backing which industry in Britain has given to metric, along with the extra impetus towards metric that came within Britain from its push in the 1970s to become a member of the EU. For the most part, industry in the United States has yet to make a real push for conversion.

A more accurate assertion about the United States would be that its consumers have held on much longer to the measuring units with which they have long been accustomed. So far, this tenacity has been accommodated by the countless businesses that target United States consumers. Meanwhile, Congress has not stepped up to the political challenge of setting a mandatory date for conversion to metric. Moreover, Americans have yet to be told that the time has come to embrace the same units of measurement that have now become customary for most of the other populations on the planet. Both the traditional U.S. system and the metric system show up in our daily lives.

Groceries: At the grocery store, in produce, the scale to weigh a bunch of bananas typically shows both pounds and kilograms – pounds, in larger numbers, on a more prominent outer ring, and kilograms, in smaller numbers, on a smaller concentric ring. In canned fruits and vegetables, the labels show quantity first in U.S. units and then also in metric units. The same is true in dairy, bottled soft drinks, and cosmetics. But at the meat counter, the butcher's scale likely shows the weight of a roast only in pounds, not kilograms.

Recipes: In the kitchen, essentially all of the recipes found in books, on food packages, or on the internet, like all of those on grandmother's index cards, are couched in measuring units of the U.S. system. Measuring cups and spoons are marked to follow the recipes, but many now also indicate metric equivalents. Finding a recipe in metric units would be highly unusual.

Automobiles: The speedometer reads more prominently in miles per hour, but it also can be read in kilometers per hour. Road signs show speed limits in miles per hour and distances in miles (efforts in the 1970s to convert signs to metric units were received, it is fair to say, with very little enthusiasm). At the filling station, the pump displays gallons sold, not liters. When the vehicle was manufactured, the assembly line employees, in a carefully organized process, pieced it together with components from a multitude of suppliers, both domestic and foreign, using parts sized in both U.S. and metric units. When the car is taken to a repair garage, the mechanic might very well need to use not only so-called "standard" wrenches (sized in U.S. units) but also metric tools.

Measurements

Drugs and cosmetics: At the pharmacy, as mentioned before, prescriptions are both written and filled using the metric system. But over-the-counter drugs and cosmetics on the shelves, if sold by weight or volume, must carry labels that use the U.S. system. Metric might also be stated.

Clothing: Clothing is almost always sized in inches, except items that can be marked simply S (small), M (medium) or L (large).

School: In the neighborhood school, science classes usually require students to be familiar with units in both the U.S and the metric systems. Indeed, the Common Core Curriculum, which like metric has had a very mixed reception across the country, calls for students to be taught both. Students will likely encounter both when they take a test.

Sports: On the sports fields, the baseball diamond will be laid out with 90 feet between the bases and 60 feet, 6 inches from the pitcher's mound to home plate. The football field will be 100 yards in length between the end zones and 160 feet in width. In the gym, the basketball court will measure 94 feet by 50 feet. All of these dimensions are in U.S. units, not metric. But at the full-sized Olympic pool, the length will be 50 meters and the width of each lane 2.5 meters. Likewise, in official track and field events records are kept only in metric units. On the tennis courts, the overall length will be 78 feet (or 23.774 meters, as the International Tennis Federation would prefer) and the overall width (for doubles) will be 36 feet (or 10.973 meters, ITF's alternative spec).

Spirits: In the wine store, a standard wine bottle label shows 750 ml (0.750 liters), while the half-bottle, the demi, shows 375 ml (0.375 liter) and the larger one, the magnum, 1.5 liter. U.S. units are not shown. But in the stacks of six-packs, domestic beers come in standard 12-ounce cans or bottles. The imports often follow that labeling, but sometimes they appear in 0.33-liter cans or bottles (about 7 percent smaller than the domestic 12-ounce portion) or in larger bottles marked in liters. The liquor bottles all show their contents in liters.

Workplaces: In U.S. offices, in those functions where paper rather than computer is still the medium, the supply is sized in U.S. units, most typically 8 ½ x 11 inches. That compares with the standard international size, so-called A4 stock, which measures 210 x 297 millimeters (8.27 x 11.7 inches). Office copiers and printers that crank paper through their workings are marketed internationally and, hence, designed to accommodate the stock that is fed into them. With the advent of digital technology, our factories have gained much greater flexibility. In the 19th century, machine tool manufacturers actively opposed conversion to metric. Now, in the 21st century, automated equipment, including robots, can be told digitally what to do, and they care not if those instructions came from specifications expressed in U.S. units or metric units.

So, in the United States both the U.S. system and the metric system find their places in everyday life. Americans, in their measurements, are living a dual existence. This duality has two principal sources. Legacy, of course, has been the principal source. Americans have attached themselves firmly to the units that the nation's settlers brought with them over past centuries. Those units – the foot, pound and gallon, and their subdivisions and multiples – have attained, through more than three centuries of national growth, both nationwide uniformity and prominence. The reason for sticking with them is obvious. Setting science aside, if it is not broken, why fix it?

Federal laws have also had their influences, with Congress generally accommodating the entrenchment of the U.S. system. The Fair Packaging and Labeling Act requires, with some exceptions, that content measurements appear in both U.S. and metric units on prepackaged consumer commodities. That hits most items on the grocery shelves, but not the fresh fruits and vegetables and not the unpackaged meats. The Food, Drug and Cosmetic Act reaches labeling of cosmetics and drugs, generally requiring that U.S units be used while allowing, but not requiring, that metric units also be stated. The Alcohol Administration Act mandates that wine, malt beverages and liquor labels show quantities in liters, but beers that do not constitute malt beverages fall into the dual labeling requirements of the Fair Packaging law. Lawyers advising food and beverage producers have many rules to keep straight.

The NIST in many ways functions as a bridge between federal and state authorities. For states, the NIST often provides guidance on what federal law requires while it fosters uniformity across state lines. On the federal front, the NIST generally acts a cheerleader for the metric system, seeking where it can to bring the United States into line with the widespread acceptance of that system within other nations.

Even more so than in other countries, the general public in the United States has been resistant to the notion of leaving behind their traditional units of measurement and moving on into the metric system. Apparently, the case still needs to be made that the U.S. system is broken enough to deserve a fix. Perhaps, just perhaps, over time the public will yield at least enough of its self-professed sense of exceptionalism to bring its measurements into harmony with the rest of the globe.

China Comes Along

Much of what is now known about measurements in ancient China was uncovered in the last half of the 20th century. The

Communist government, after taking control in 1949, gave importance to discovering more about that nation's history. As a part of that effort, starting in 1976 the National Institute of Metrology (NIM, the Chinese government's scientific research institute) had a small group of researchers engaged in finding records and artifacts which would tell the story of China's metrology. The results of that research were published in books and articles, but then, about 20 years after its launch, the investigating group was disbanded.

Not surprisingly, the history of measurements in ancient China is similar to that in the West. From about 2000 BCE onward, the territory, or portions of it, which is now generally deemed to be the nation of China had a succession of governing dynasties, some lasting for several hundred years and others for only a decade or so. Each dynasty had its own approach. The Zhou Dynasty (1046 – 256 BCE) bestowed powers upon aristocratic families as lords of their own domains, much as kings and queens did in Europe during the Middle Ages. Various lords exercised their powers by dictating their own measuring systems. Then, on the heels of the Zhou Dynasty, in 221 BCE came the short-lived Qin Dynasty (221 – 206 BCE), which took it upon itself to impose centralized administration. Among other steps affecting the entire population, it fixed universal standards of measurement.

As was the case in other early cultures around the globe, in China parts of the human body, such as the hand, fingers and feet, were the earliest tools of measurement. Over time, however, units of measurement were given more certainty by representations on sticks, lines or bulks of millet seeds and other physical objects which could be widely reproduced.

Within the courts of government, successive dynasties, in a pursuit of spiritual harmony, put emphasis upon using correct measurements for all things that would reflect upon their right to rule. This influenced their court rituals, including garments worn, carvings displayed and bells rung. Outside the courts, everyday life in China went on but not without being affected by the changes that ruling dynasties were requiring as demonstrations of their authority. Measurements in the marketplace needed to bear some relationship with those in the dynastic courts. As they were doing elsewhere over the centuries, measurements in China evolved and, with the growth of the economy, became increasingly uniform across dynastic territory. This was happening even while Chinese culture remained essentially free of outside influences.

It was not until the 19th century that traditional measurements in China confronted challenge. The Opium Wars in mid-century put major port cities under the influence of foreign traders, who brought their own units of measurement as replacements for those long used by inhabitants. In some ports the foreigners took control of the customs offices. While less true outside the coastal cities, there was a sense that long-standing Chinese traditions were no longer operative.

In the first years of the 20th century, the empress dowager of the Qing Dynasty (1644 – 1912), seeking ways to bolster the fading legitimacy of her rule, instructed the Chinese ambassador to Paris to consult with the BIPM about conversion to the metric system. In 1908 that dynasty redefined traditional weights and measures by stipulating ratios between its units and their most approximate counterparts in the metric system. That step towards reform kept the names of traditional units but assigned to them somewhat different sizes.

Political revolution could not be halted. The Qing Dynasty was replaced in 1912 by a government that called itself the Republic of

China, under the leadership of a provisional president, Sun Yat-sen. But concern about the lack of uniformity in Chinese measurements continued. New representatives were sent to consult with the BIPM. A Bureau of Measurement was created in China to improve the existing system. Ongoing social upheaval, however, slowed progress.

When Sun Yat-sen died in 1925, Chiang Kai-shek took his place. The push for reform continued. In 1929, Chiang Kai-shek issued a law that retained traditional measurements for internal use but adopted the metric system for official transactions. The 1930s and 1940s were years of tumult in China. Japanese occupation, a power struggle between Chiang Kai-shek's government and Mao Zedong's Communist uprising, and then World War II and the destruction it brought not only to China but all of Southeast Asia left little space for pursuing reform of measurements, or of anything else. The turmoil did not begin to subside until 1949, when, with the global war over, Chiang and his nationalist Government were driven from the mainland and Mao came to power as the head of The People's Republic of China.

Under Communist control, China rejected the politics and lifestyles of the West but, and this is a big "but", it recognized that the metric system would be a critical component of bringing Chinese science and technology up to global standards. Still, internal political strife made conversion to metric an on-again, off-again priority. The Cultural Revolution, launched by Mao in 1966, brought death, imprisonment and anarchy to millions. This was the story of China until 1976, when the so-called Gang of Four was purged and then, only a few months later, Mao died. China could once again try to put itself on even keel. China became a member of the group of nations participating in the work of the BIPM, and full conversion to the metric system was legislated in 1985.

Like the United States, modern China now lives with two systems of measurement, a traditional system and a metric system. And again like the United States, those two systems are both fundamentally based upon BIPM metric units (recall the Mendenhall Order of 1893 which redefined traditional units in the U.S. system by reference to units in the metric system).

A difference between China and the United States is that in China the sizes of units in the traditional system of measurement, like those in the metric system, have been aligned with the decimal numbering system. (Archaeologists in China have discovered rulers for measuring with decimal subdivisions. They date these to the second millennium BCE.) For example, the *chi*, a traditional Chinese measurement which approximates the foot, has subdivisions of tenths (a *cún*), hundredths (a *fēn*), and so forth.[40] If simplicity is a goal, that method trumps subdividing the foot into inches, half-inches and so forth.

As mentioned, China has a national laboratory, the NIM, which functions for that nation in much the same role that the NIST performs in the United States. First established by the Communist government in 1955, the NIM now occupies two facilities. The primary one, new in 2009, is about an hour's drive northwest of Beijing. The smaller one is within Beijing. Together, they have a staff of about 1,000.[41]

In 1977, China became a participating member of CGPM, the international body that oversees the BIPM. Since 1949, when the Communist government under the leadership of Mao Zedong took control, the country has been endeavoring not only to unify its internal standards of measurement but also to make a full conversion to the metric system. Even though metric was recognized, at the outset of Communist control, to be an influence of the West which, for that reason alone, would have otherwise been summarily rejected, the authorities saw full conversion to metric to be an essential key to national progress. As history now has it, conversion to metric, like many other goals, was hampered by the political and social turmoil that continued in China up through the end of the Cultural Revolution in 1976. It was only then, in the late 1970s, that China could begin to open itself to the West. Now, decades later, the NIM's scientists are working in collaboration with those at the BIPM and other national laboratories around the globe to maintain uniform international standards of measurement and to improve existing standards.[42]

Chapter Four
CALENDARS

Flight 1230's on-time departure from Washington-Dulles, for its non-stop to Beijing, reflected careful planning. Passengers had to purchase their tickets and reserve their seats. The maintenance crews had to check essential systems on the aircraft and complete all necessary repairs. Then as flight time approached, the fueling crew had to fill the tanks, the cleaning crew had to spruce up the cabin, the baggage crew had to load the lower hold, and the catering crew had to stock the galleys. And, of course, both the flight crew and the passengers had to get themselves to the gate and on the plane. And then after all of that, the tower crew had to coordinate Flight 1230 with other traffic on the tarmac and taxiways to sequence the Beijing-bound flight onto the operative runway. When Flight 1230 touched down in Beijing, a very similar sequence took place, first to unload the aircraft and then to make it ready for its return flight.

The passenger who occupied seat 22A on Flight 1230 was a research scientist at the National Institutes of Health. He was traveling to Beijing and then on to Chengdu in southwestern China to join his family for the Chinese New

Year. Departing Washington-Dulles in mid-afternoon and arriving in Beijing late the next day (local Beijing time), he would have an overnight stay at an airport hotel and then make an early morning connection to Chengdu. Not having returned home since first coming to the United States seven years before, he was very much looking forward to joining his parents and other relatives when they would gather for a traditional family dinner to celebrate the Chinese New Year.

Many people, many duties, a great deal of coordination. Glitches happened, but on most days the important things all proceeded on schedule.

Time Rolls On

For much, much longer than Ole Man River, time just keeps rolling on. Humans are learning not only how to track it but, of equal importance, how to fit their individual lives into the unstoppable flow. This is the *why* of both calendars and clocks. The few philosophically inclined souls who have taken a moment to ponder the unstoppable flow have sometimes told us we are all living under a tyranny of time. They have a good point. The notations we make on our calendars and the frequent glances we give to our clocks can be put into context by comparing our lives today with those in the past.

In ancient times, when the focus for many was growing crops for their own meals, it was sufficient simply to observe the change of seasons and the rise of rivers from which irrigation could be drawn. Coming forward by centuries, up through the Middle Ages, across Europe it was principally the church clerics who, in fulfillment of their responsibility for the timely observance of religious rituals, gave close attention to the passage of time. Among the citizenry, letters were seldom dated and, when they were, it was by the year of the sitting king's reign, not the numerical sequence we now follow. As late as the 15th century, whenever someone's age had to be formally determined in England, the local sheriff would often need to call a jury of local residents who knew the person. Parish birth registers did not become official until 1538.

Almost five centuries later now, time may very well be an overarching tyranny in our lives, but at least it is a tyranny shared around the globe and, for that reason, serving as good evidence of the ongoing unification of Earth's inhabitants. For all practical purposes, the entire world shares a single calendar (years, months, weeks and days) coupled to a single clock (hours, minutes, and seconds). A full appreciation of this global uniformity comes out of an understanding of how many different timekeeping traditions from diverse cultures had to be melded together over centuries past.

Timekeeping is based upon two regularities found in nature: first, the Earth orbits the Sun in an essential regularity (a solar year); and second, the Earth revolves on its own polar axis in an essential regularity (a day). (As will be seen in the next chapter about clocks, these two regularities are not perfectly regular when observed with the precision of modern instruments. A new basis for precision in timekeeping has been found inside an atom.)

In some ancient cultures, these two regularities were overlooked. Instead of measuring a year by the Earth's orbit, it was measured by observing the changing phases of the Moon in its orbit around the Earth. Twelve completed orbits of the Moon were deemed to be a year (the lunar year). But as it became increasingly evident that the orbit of the Moon around the Earth is not synchronized with the orbit of the Earth around the Sun (or as once understood, the Sun around the Earth), it became necessary to add days to the lunar year before moving forward to the succeeding solar year. Over time, as cultures assimilated and astronomical observations advanced, the lunar year gave way, at least for civic purposes, and the solar year came to the fore.

It is worth noting that once the regularity of the Earth's orbit and that of the Earth's rotation are taken together to be the foundation of timekeeping, all units of timekeeping that have been built on top of these two natural phenomena are human-made structures, not rules of nature. The divisions of a year into months, a month into weeks, a week into days, a day into hours and so forth are just timekeeping conventions. These conventions serve their purposes because, but only because, humans have agreed to live their lives together by following them. Their usefulness has always been and still is proportionate to their usage.

Fixing our minds now on the regularity of the Earth's orbit around the Sun and the regularity of the Earth's rotation around its own polar axis, in this chapter about calendars and the next about clocks we can look at how the global community has come to the timekeeping practices it now uses to coordinate its affairs.

This is nothing less than a tale about the multiple and often conflicting concerns of humankind. To identify only some of these concerns, the most important ones over centuries have included growing crops (farmers), expanding commerce (merchants), honoring religious beliefs (clergy), advancing the sciences (astronomers and others), navigating land and sea (explorers, mariners), promoting, expanding and protecting nations (diplomats and armies), and adhering to established customs (diverse cultures).

Gregorian Calendar

The global community for all practical purposes now uses what is most commonly called the "Gregorian calendar." It is so named because Pope Gregory XIII, in 1582, acting on the advice of a special commission, decreed two changes in the Justinian calendar which had come from the Romans.

First, the Pope advanced October 4, 1582, to make it October 15, thereby correcting for an eleven-day loss that had accumulated due to the Romans having imprecisely deemed the length of each year to be 365.25 days. The Pope had been advised that the Romans had put an extra 11 minutes into each year (one day every 128 years). The count 365.25 was a small fraction of a day too long and for that reason more "time" had actually elapsed than had been tallied on the Justinian calendar.

To avoid future accumulations of that error, Pope Gregory XIII directed that the leap year (adding February 29, to account for the fractional 0.25 day) be omitted in all future centenary years (years 1700, 1800 and so forth) except those that are multiples of 400 (years 1600, 2000, and so forth). With this seemingly small adjustment, years would pass at a slightly more rapid pace in future centuries. (But all still was not made perfect. The Gregorian calendar, with slightly fewer leap days, is nevertheless accumulating an error of about one day every 3,236 years. The Gregorian calendar is about 25 times more accurate than the Justinian calendar it replaced, but it is not precise.)

Second, and unrelated to correcting the measurement of the year, Pope Gregory XIII decreed that the calendar year would begin on January 1. This brought the starting date into line with that which had been adopted by the Romans in 153 BCE to make the start of each calendar year coincide with the start of one-year terms for Rome's elected consuls. Early Christians, rejecting this Roman convention as a pagan practice, had subsequently adopted a starting date of March 25, thereby commemorating the Day of Annunciation.

Venice and other localities had used yet other starting dates, some of which continued into the Middle Ages. In 1582 the Pope, upon introducing the Gregorian calendar, sought to eliminate these inconsistencies in the starting date. Like his adjustment of the length of a year, his change of the starting date was not widely accepted for some time to come.

Protestant populations in post-Reformation Europe were slow to adopt the Gregorian calendar, even though many Protestant astronomers, notably Tycho Brahe and Johannes Kepler, approved its changes. By 1700 most Protestant states had adopted it, but the new calendar did not become official in Britain and its colonies, including those in North America, until 1752. It took the Orthodox Church in Greece, Romania and Russia until 1923 to embrace it.

A notable diversion from the Gregorian calendar occurred in France in 1792, after King Louis XVI had been deposed. The French Parliament decreed that "Year I" would begin on what would otherwise have been September 22, 1792, the day the new Republic had been proclaimed. Moreover, each year would have twelve months with a new French name, each month would consist of three weeks each lasting ten days, and then a festival of five additional days would conclude each year. This new all-French calendar, which the American statesman John Adams called a "course vulgarity", had a short life. On Napoleon's decision, France reverted to the Gregorian calendar on January 1, 1806.

Years on the Gregorian calendar, of course, are distinguished from each other by sequential numbers, counting from the year that Christians deemed to be the year when Jesus had been born. Years subsequent to that Christian event are often designated AD, which is the abbreviation for the Latin phrase "*anno Domini*," meaning "in the year of the Lord." Years preceding the birth of Christ are designated BC, meaning "before Christ." This numbering method has been traced back to a set of tables prepared by a Roman monk in 525 AD. It was used 200 years later in an English treatise, but its usage then lapsed until the 15th century. In the mid-19th century, Jewish academics introduced the alternative designations CE ("common era") and BCE ("before common era"), and by the late 20th century many academics and publishers were using these alternatives to show sensitivity to non-Christians.

Backstory

The backstory for our Gregorian calendar can be traced to Mesopotamia and Egypt. Obviously, earthly inhabitants have always had reason to keep track of time for their own purposes, not the least of which in agrarian economies has been when to plant and harvest crops. It was, however, the populations dwelling along the Euphrates and the Nile during the two millennia before the common era who initiated the methods of timekeeping which would ultimately provide the basis for the Gregorian calendar and other contemporary timekeeping tools.

It was not until the work of the Polish astronomer Copernicus, in the 16th century, that the Earth was recognized to be a planet revolving around the Sun, not the Sun around the Earth. Even then, it took much time and much turmoil, most notably for Galileo in the early 17th century, for Catholic doctrine and humankind in general to put the Earth rightfully into its solar orbit. So, throughout the two thousand years that it took the Gregorian calendar to evolve from ancient methods of timekeeping, the Earth was erroneously deemed to be at the center of the universe with the Sun circling around it. The ancients and medievalists had it all wrong, yet they were able to devise a calendar that, with refinements, has withstood the test of time, literally.

On the Euphrates, the Babylonians used a lunar calendar of twelve months, but because the Moon circles the Earth more than twelve times during a solar year they had to add a thirteenth month from time to time to keep the two cycles more or less synchronized.

As a side note, Babylonian astronomers are credited for their early work with the zodiac, the celestial belt that follows the path of the Sun across the sky. Due to the tilt of the Earth's polar axis, the celestial belt is askew the terrestrial equator. Around 700 BCE, ancient astronomers divided the zodiac into 12 equal segments of 30 degrees each, consistent with the Babylonian practice of using 60 as the base number and 360 degrees as a full circle. (The Babylonian division of the circle into 360 degrees has become, of course, an axiom of modern mathematics. Could it not just as well have been 10 degrees in a full circumference? Or 100? Or 1,000? The number 360 is critically important now, but only because it has been so universally accepted as both a tool of mathematics and a frame of thought.)

In Egypt, life was largely governed by the annual rise of water in the Nile. A so-called civil calendar was based upon the average time between these once-a year events, and that average was divided among 12 months of 30 days each, with an additional five days tagged to the end of each year. When it was later recognized that the astronomical year, the solar year, was a fraction of a day longer than these 365 days on its civil calendar, the Egyptians began also to use a second calendar based upon tracking the star now identified as Sirius. This astronomically based calendar more closely, but not perfectly, kept pace with the seasons as they were being observed. Separately, the Egyptians had a lunar calendar to regulate festivals by the phases of the moon.

The Jews, like the Babylonians, based their calendar on the lunar year and are thought to have begun the practice of successively numbering the years. For obvious reasons, the Jewish numbering used a starting point long before (indeed, 3,761 years before) the starting point later imposed by Christians, the one now widely used with the Gregorian calendar.

In Greece, each city-state had its own calendar, years were identified by the names of local officials, and it was those local officials who decided when to add a thirteenth month to the year to keep the lunar cycle better coordinated with the solar cycle.

As the Romans extended their geographic reach, the numerous and diverse local date-tracking practices they encountered were at first left to be within each locality, but later the variations among the localities were seen to bring too much confusion. Moreover, some of the local practices were inaccurate. In 45 BCE, Julius Caesar came to grips with these problems by introducing what came to be known as the Justinian calendar. Unlike the lunar calendars it superseded, this new calendar was based upon a solar year of 365.25 days. It took account of that troublesome fractional day of 0.25 by establishing a leap year with one additional whole day regularly placed in every fourth year.

It was this Julian calendar which in 1582, some 1600 years later, came under scrutiny by Pope Gregory XIII. In particular, it was this calendar's accumulating inaccuracy, making a full day correction in *every* fourth year (the leap year), that resulted in the Pope's declaration that the Gregorian calendar should henceforth be used in its stead. As we now know, that papal declaration has taken deep root.

Chinese Calendar

Recall that Flight 1230 out of Washington-Dulles made a nonstop trip to Beijing. The passengers had to reset their watches for the time change, but other than fixing into their heads that they had gained a partial day due to crossing the international date line they were able to continue using the same calendar they had used to board the flight in the United States.

In all likelihood, the Gregorian calendar was introduced in China by the Jesuit missionaries sometime after Pope Gregory XIII decreed its adoption in 1582. Their goal was to impart Christianity to that far-distant population. The calendar they took with them probably found increasing familiarity as China's contacts with Europe became more frequent, especially during the 19th century when British traders were implanting themselves on the Chinese mainland. The Gregorian calendar, however, was not officially adopted until 1912 (some sources put this in 1929, when the Nationalist Government issued a decree) and it did not come into widespread usage until 1949, when Mao Zedong, as head of the Communist Party, ordered that the year be in accord with the Gregorian. The Gregorian calendar is now used across China for civil purposes, but the people there and in Chinese communities around the globe also use the traditional Chinese calendar for determining dates of festivals, such as the Chinese New Year, and favored dates for special occasions, such as weddings.

4.1 Traditional Chinese calendar with Gregorian years added

That traditional Chinese calendar is lunisolar, which means it is based upon both lunar and solar observations. Years are not numerically sequenced. Instead, they are named according to their order in a sequence of 12 animals. Within each year, each month begins on the date of a new moon. Most years include 12 months, but to account for the nonsynchronous orbits of the Sun and the Moon some years bring 13 months. The calendar is often presented as a circular tabulation with 12 sectors, each sector representing the animal whose name is attached to a period of five years, the sets themselves separated from each other by 12 years, in the 60-year cycle.

Other Calendars

China is not alone in its continuing use of a traditional calendar for purposes other than civic affairs. The Jewish calendar, lunisolar like the Chinese one, is used to determine dates for religious observances. Within the nation of Israel, the traditional calendar continues to be the official one even for civic purposes, while the Gregorian calendar is gaining ground in its usage. The Islamic calendar, a lunar one, has a holding on religious practices within Muslim communities, as does the Hindu calendar for adherents to that religious belief. Each of these calendars, Jewish, Islamic and Hindu, as can be expected, has its own starting date tied to the creed of its faith.

Turning to Amerindian civilizations, the Mayan culture in Central America during the first millennium CE had three calendars which were used simultaneously. One was the so-called "Long Count" calendar, which some believe might have been used to track time since the creation of the world. The second was the "sacred almanac" (in Mayan, the *Tzolk*), tracking the religious year of 260 days, and the third was a civil calendar, tracking the solar year. With only spotty records to guide them, scholars have long puzzled how the Mayans coordinated the three calendars.

We can feel a sense of relief in now having the Gregorian calendar accepted worldwide as a common reference for coordinating civic affairs. It is well and good that some cultures have also retained their own calendars for special purposes, but having a single calendar for civic purposes makes it possible for the world population as a whole to move on together as our lives become increasingly intertwined.

Months, Weeks and Days

Recall that the Babylonians and Egyptians both divided their lunar years into 12 months, adding days at the end of each 12-month period, or alternatively adding a 13th month to some years, in an effort to stay in step with their observation of 365 days, plus a fraction, in the solar year. That division by 12 made its way into the Justinian calendar introduced by Julius Caesar in 45 BCE. The Justinian calendar took direct account of the solar year by using six months of 30 days each in alteration with five months of 31 days each, plus treating February as an outlier with 29 days in three successive years and 30 days in the fourth year (the leap year). The total Justinian year thus had 365 days, except 366 in each leap year. A new year started on March 1.

The months on the Justinian calendar were juggled in 7 BCE after Augustus Caesar had succeeded his great-uncle Julius. Without renaming the calendar, Augustus renamed the sixth month (counting from March as the first month) to be Augustus (no pause to wonder why) and gave it 31 days, thereby matching the number of days in July (which had earlier been renamed in honor of Uncle Julius). To give his namesake month this equal dignity, Augustus took a day out of February and, to avoid leaving three successive months of

31 days (July, August and September), he reduced September to 30. To complete his realignment, Augustus increased both October and December to 31 and dropped November back to 30. Net result: 365 days still in non-leap years, 366 in leap years. This reworking of the months, two thousand years ago, put us where we still are on the Gregorian calendar.

The Romans also put the names on our months. They took them from deities (January, March, May and June), an annual festival (February), the opening of buds (April), two emperors (July and August), and numbers (the Roman year started with March, so month seven was *Sept*ember, eight was *Octo*ber, nine was *Nov*ember and ten was *Dec*ember).

Along with their early influence on the counting of months, the Babylonians set the pattern for what we call the week. They gave attention to the seven-day periods associated with phases of the moon and, to appease their gods, prohibited certain behaviors on the seventh day (the "evil day"). This practice influenced the Jews and early Christians and is deemed to be the origin of the seven-day week, as well as the historical reason behind restrictions that some communities place on activities on the seventh day, the "Sabbath" (in the Jewish faith this is Saturday, in the Christian faith Sunday). The concept of a week, of course, is nothing more than an accepted convention. We commonly organize our lives around the notion of a week only because our fellow global citizens do the same.

Like the seven dwarfs, in many languages the seven days of the week have their own names. In English, the names Sunday, Monday and so forth have all been traced back to Roman origins, which are thought to have been derived from the classical names of planets in Hellenistic astrology. For example, when the Sun was thought to be a planet circling the Earth with other planets, the Roman name for

Sunday was *dies Solis*, meaning "Sun's day." At the opposite end of the week, the Roman name for Saturday was *dies Saturni*, an obvious take on the planet Saturn.

In some languages, however, days are numbered, not named. And in these cases the day considered the first day can vary, with the result that different numbers attach to different days, depending upon the local culture.

In contrast with the week, the concept of a day is more than a convention. It embodies a regularity in nature, the time it takes the Earth to do a full rotation on its polar axis. Through the centuries a day as a measure of time has always been a constant for earthly inhabitants, even though different cultures have used different markers to determine the end of one day and the start of the next. Sunset was the marker for Babylonians, Jews and Muslims, and that legacy continues for some religious purposes. Dawn was the marker for much of Europe until the 14th century brought the striking clock. Noon was the start of a new day for astronomical purposes until as late as 1925. In modern times, midnight is the pivot point from one day to the next, a practice once followed by the Romans to eliminate the uncertainty which is inherent in the variable length of total daylight hours.

International Date Line

While Pope Gregory XIII in the late 16th century was working wrinkles out of his namesake calendar, global explorers were discovering it has a big rupture.

In the 21st century, the westbound passengers on Flight 1230 from Washington-Dulles to Beijing were aware that they would cross the international date line somewhere over the Pacific. Five hundred years earlier, in the 16th century, the daredevil mariners who were first to circumnavigate the globe were wholly unaware of that subtlety, at least until they had returned to their homeports.

Ferdinand Magellan, in command of a five-ship fleet, set sail westward from Spain in 1519. He discovered the Strait of Magellan around the tip of South America and proceeded across the ocean now called the Pacific, arriving in the Philippines almost 18 months into the voyage. Here, Magellan was killed by a poison arrow in a local battle. One of the two remaining ships in the fleet, after loading a cargo of spices, turned back eastward but later perished. The second one, the Vittoria, similarly loaded, continued westward, rounded the Cape of Good Hope and, three years after the start of the expedition, made it back to its homeport of Seville in September 1522.

Having used the Sun to tally the days throughout the long voyage, when the Vittoria reached Cape Verde off the west coast of Africa and its crew inquired about the day of the week, the sailors were surprised to learn the day was a Thursday, not a Wednesday as they had reckoned. Their arduous adventure brought to their attention the conclusion of a thought experiment which had been explained, but not tested, almost two hundred years earlier.

A French bishop and polymath in the 14th century had theorized about two ships setting out in opposite directions to circumnavigate the globe. When the two vessels both returned to the homeport at the same time, the bishop concluded that the eastbound craft would show its arrival one day beyond the local time while the westbound craft would show one day behind.

Our planet really is a sphere, just as the ancient Greeks had postulated! It rotates eastward, into the sunlight. For travelers going eastward 15 degrees in longitude (roughly, the distance from Chicago to New York), the duration of one day shrinks about four minutes. Likewise, for those traveling westward 15 degrees, one day expands by about four minutes. Those time differences accumulate the farther one travels. After a full circumnavigation of the globe, the entire 360 degrees, the accumulation is 1440 minutes, which equates to 24 hours or one full day. A thought experiment, indeed. It took real life experience to drive this understanding eventually into widespread awareness.

In the 16th century, the awareness spread slowly. When ships from Spain, sailing westward, and those from Portugal, sailing eastward, came together in Asian ports, they found that the days marked on their calendars did not coincide, one being a day ahead and the other a day behind in the foreign port. The first Englishman to circumnavigate the globe was Sir Francis Drake. Having sailed in a westward direction, upon his arrival back in England in 1580 Drake discovered that he had lost a day. By his tally the day was Sunday, but in England the day was Monday. As awareness of this subtlety spread, the custom developed among mariners during the 17th and 18th centuries to change the date aboard ship before returning to homeport. In the first half of the 19th century it became increasingly common to make this change upon crossing the 180 degree meridian in the Pacific.

Similar issues arose, of course, for explorers on land. Russian explorers migrated into Alaska and then worked their way southward as far as San Francisco Bay. They used what was known as Eastern dating (adhering to days counted by eastbound travelers), and by doing so their days did not align with those of indigenous groups they encountered. The conflict was resolved in the late 1860s when the United States purchased Russian interests in Alaska and what was then known as American dating (adhering to days counted by westbound travelers) covered the entire North American continent.

Thus came to be what we now call the international date line:

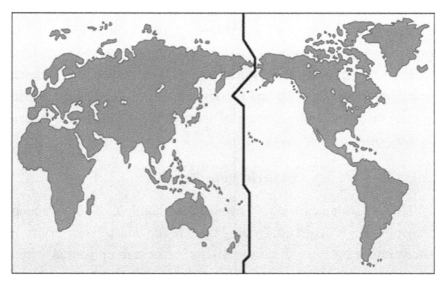

4.2 International Date Line   Some island nations in the Pacific on one side or the other of this line have chosen to set the dates on their calendars as if they were on the other side. This means the actual line has some kinks.

This international date line has no official status. Sovereign governments are entitled to fix unilaterally their own date lines and time zones within their own lands and adjacent territorial waters, and they do so according to their own interests. As a consequence, when these sovereign choices are taken into account the line, while generally tracking the 180 degree meridian, takes several zigs and zags up and down the Pacific, leaving some sovereign island nations on what logically would be the wrong side. Those apparent misplacements are important, of course, to those who reside in or travel to these island nations, but they matter not to those who, like the passengers on Flight 1230, cross the Pacific nonstop.

In 1917, a nautical date line was established by international agreement at a London conference. Like the unofficial international date line, this more official line tracks the 180 degree meridian, except where it is interrupted by territorial waters. All naval ships and many non-naval ships are required to use the nautical date line and the widely recognized time zones, running east and west from Greenwich, when not in territorial waters. In practice, that requirement is honored only for radio communications. On board, a ship may use a timekeeping system of its own choosing for meals and similar purposes.

Worldwide Usage

The Gregorian calendar is an excellent example of global acceptance of a single standard. It came to us in the late 16th century from a declaration by Pope Gregory XIII, after he had consulted with a special commission about correcting errors that had been accumulating under the Justinian calendar inherited from the Romans. Other calendars with their own histories are used for religious and other purposes by various groups, such as the Chinese, the Muslims, and the Israelites, but even these groups have accepted the Gregorian to manage their civic affairs. The divisions on the Gregorian into months and weeks each has its own historical explanation, but the count of 365 days, or 366 in a leap year, is a measure of the time it takes for the Earth to circle the Sun and the duration of one day is a measure of the time it takes for the Earth to make one revolution on its own axis. Neither the solar path around the Sun, a phenomenon of nature, nor the revolution of our planet on its own axis, another phenomenon of nature, gives humankind any other reasonable choice for measuring a year or a day.

Our ancestors paid attention to natural phenomena when they were working out how to keep track of the passage of years and days. Different cultures long ago had their own understandings, but eventually scientific observations, specifically the Earth's revolution around the Sun and its rotation around its own axis, prevailed. The result is that on a global scale essentially all of us are using a single calendar to manage our lives together on one planet.

Chapter Five
CLOCKS

When Flight 1230 passengers purchased their tickets, they saw on the schedule that the flight would depart Washington-Dulles at 2:15 PM and arrive in Beijing at 5:30 PM the next day, with a flight time of 14 hours 15 minutes. Knowing that flight schedules customarily use local times, most passengers saw nothing in those scheduled departure and arrival times that puzzled them. During the first hours of the flight going west, hours would slip back as time zones were crossed, but when the international date line was crossed 24 hours would be immediately gained. Then, continuing westward, hours would again slip back. At touchdown, the net time in flight would be 14 hours, 15 minutes.

At Washington-Dulles, 233 passengers boarded Flight 1230 before the door was closed. As is always the case, a few just barely made it. But two passengers were no-shows. One of the two had overslept after attending a festive wedding the night before, and he was still in a taxi racing to Dulles when Flight 1230 pushed back from the gate. The other, a resident of Toronto, had been the first passenger to board her connecting flight from Canada at mid-morning but that *flight had then been delayed by fog at Toronto Pearson field. She missed her Dulles connection to Beijing by 20 minutes.*

There can be little doubt that both of the no-shows had their attention glued to their watches, hoping upon hope that they would still make it to Flight 1230 before the door would close. Likewise, there can be little doubt that in the cockpit both the pilot and the co-pilot had their attention fixed on the instrument panel in front of them, including the digital flight clock that would tell them when they could order the door to be closed for a timely departure. All eyes were on clocks!

When planning their trips, the passengers on Flight 1230, especially those making a connection at Washington-Dulles after arriving from a foreign country, would have needed to know not only the Washington, D.C. time zone but also the use of daylight saving time, its beginning and ending dates, in both Washington and Beijing. And for those passengers needing to make a connection in Beijing to another destination, the need to know about China's use of daylight saving time was a compounding factor. The brief notation "All times are local," typically found on published flight schedules, is hardly enough.

Timekeeping Conventions

Our clocks divide our days into hours, hours into minutes, and in many cases minutes into seconds. Our most precise clocks, of course, continue this dividing and subdividing by breaking seconds into tenths, tenths into hundredths and so on. These timekeeping conventions, uniformly honored on a worldwide basis, play a critical role in holding the global population together. We too much take them for granted. They deserve a day in the sun.

> It is a curiosity that the second, as a unit of time, is subdivided into smaller units on a metric scale (tenths of a second, hundredths, and so forth), rather than the ancient base of 60 (for example, 60 subdivisions in a full second, then 60 subdivisions in each of those, and so forth). Only scant explanation has been found. It appears that the conventions of hour, minute and second were deeply entrenched when it first became possible to measure time in units smaller than the second. The effort both to overturn centuries of habit and to replace countless clock mechanisms and faces was likely judged not to be worthwhile.

Backstory

The convention of dividing a day into 24 hours (not ten hours, not 100) has been traced to ancient Egypt. Daylight, the period from sunrise to sunset, was divided into ten equal parts and then one more part was added for morning twilight and another for evening, making 12 in total. Nighttime, the darkness, was similarly divided into 12 parts. All 24 parts were called "seasonal hours." It was not the most practical system because over the course of a year the durations of daylight and darkness would vary, causing the length of each seasonal hour to vary as well. Neither the origin nor the purpose of this early Egyptian division has been reported in available resources.

Whatever its purpose, this ancient system has survived, but only in part. The day is still thought to have 24 parts, giving us the convention we call an "hour," but the hours are all now of equal duration. In western Europe, the counting of two 12-hour segments, midnight to noon, then noon to midnight, eventually became common (counting one to 24 in each full day persisted for centuries in Italy). It has now become common in most countries around the globe.

In ancient times, only the astronomers found sufficient reason to treat the 24 parts of a day as equal intervals of time. Their calculations involved fractions, and for that purpose they needed a standardized unit that could itself be divided into standardized units. To that end, they used the equal periods of daylight and darkness on the day of an equinox as a constant measuring standard throughout the year. Then they followed the number system which came from the Sumerians and was followed by the Babylonians, applying the base 60. Hence, each of the 24 equal parts of the day, which is to say each hour, was itself divided into 60 minutes, and each of those subparts was further divided into 60 seconds. Working before the invention of clocks, early astronomers, by tracking movements of celestial bodies across the sky, were able to devise the essential tools for their science. Paradoxically, their science involved puzzling out the patterns in those same movements.

Other than the astronomers, those in the ancient world, up through Roman times, apparently found no good reason to rethink the Egyptian practice of allowing unequal intervals, over the course of a year, to the 24 parts making up a full day, daytime plus nighttime. They had no compelling need in everyday life for accurate timekeeping. Putting hours of equal intervals in all of the days, all year round, would await the advent of terrestrial timekeeping devices up to that task. Those devices would come as the need for more accurate timekeeping evolved.

Ancient Clocks

The ancients were not wholly without timekeeping devices, but until the advent of the mechanical clock in the 13th century there was no device which provided sufficient accuracy to track the minutes within an hour, much less the seconds within a minute. Most simply, a stick standing vertically in the ground would cast its shadow against markers on the ground to indicate roughly the time of day. The sundial was and is only an enhancement of this crude device. Sunshine, of course, was and is an essential component. Cloudy days shut the operation down, as did the darkness at night. The wetter weather and longer nighttime in northern climates were especially limiting. Equally simple, a burning candle or stick of incense, assuming its uniformity, would disappear incrementally at a more or less steady rate. While these medieval devices were not dependent upon sunshine, they did require much attention and, consequently, were not good substitutes.

A more interesting and more complex device was the water clock, or *clypsydra* (from Greek, meaning water thief). Its origin may have been Egypt in the 15th century BCE, but that is not certain. The water clock, with many refinements from time to time, was used both in the western world, including Greece and Rome, and in China and other parts of Asia.

In the simpler type, a regulated flow of water would fill a reservoir and markings on the inside of the reservoir would indicate the duration of that inflow. Alternatively, an empty reservoir with a small hole in its bottom would be placed in a filled vessel and allowed to take in water; the time taken to fill the once-empty reservoir would be a measure of elapsed time. These devices were often used as timers, not continuously running clocks, for the purposes of limiting testimony in courts, measuring a patient's heart rate, controlling the quantity of irrigation water a farmer could drain from a public source, or limiting a client's visit in a brothel.

But water clocks became more complicated, and more expensive, once mechanisms were added to harness the flow of water so

Clocks

that the device, more than simply serving as a timer, could show continuously running displays of time while also ringing bells, activating astronomical readings, and propelling other apparatuses. In their usage as continuously running timepieces, water clocks had to be geared to take account of changes in the length of an "hour" over the course of a year, with 12 longer hours of daylight in the summer (12 shorter ones at night) and the opposite in the winter. The Greeks and Romans had their own solutions, as did the Chinese.

One of the most notable water clocks was constructed in China late in the 11th century. Called the Su Song astronomical clock tower, it was over 30 feet tall and, with what for that time was elaborate

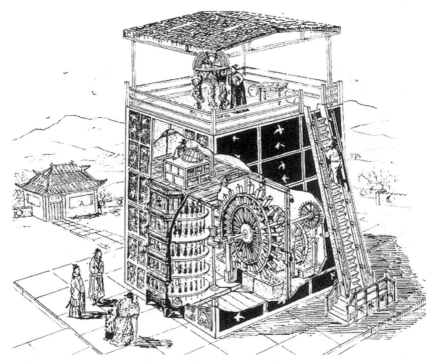

5.1 Su Song Astronomical Water Clock Tower   This structure, over 30 feet high, was constructed in China at the end of the 11th century. In addition to tracking time, it propelled astronomical displays, rang bells and activated other mechanisms.

gearing, was used to power not only a timekeeping device but also to drive globes for astronomical viewing, ring gongs, and activate other mechanisms.

By comparison with the sundial, the water clock in its most advanced design was a much more accurate timekeeping device. The complexity of that design, however, made the device expensive and, as a result, its ownership was limited to wealthy members of society, many of whom had no need for accurate timekeeping but apparently did have a need to demonstrate their wealth. The advantage of greater accuracy kept water clocks in common usage until the 17th century, when pendulum clocks took the lead.

A need for accurate timekeeping arose in the Islamic world after the inception of that religious following in the seventh century. Astronomers were called upon to determine when prayers were to be said during the course of each day. These scientists of their time deployed the astrolabe, which was an instrument designed primarily to solve a range of problems in spherical trigonometry. It has been called "the slide rule of the Middle Ages." With this device in hand to make a sighting on either the Sun or a charted star, the astronomer, by aligning its concentric circular gauges, could with some precision determine the time of day or night. The astrolabe found usage in the hands of astronomers not only in the Muslim world but elsewhere in the western world and in China.

Across Europe and other non-Islamic regions during the Middle Ages, it was equally true that the need for tracking the time of day was mostly limited to clerics responsible for the regular tolling of church bells. (The English word "clock" derives from the Latin *"clocca"* and the French *"cloche,"* meaning bell.) There was a particular need for timekeeping in medieval monasteries, where punctuality for divine services and meals was rigorously enforced. The necessary skills were most readily found among the educated individuals within the church, in general, and the monasteries, in particular.

Outside religious life, for as long as growing crops and raising animals continued to be the predominant economic activities, little attention was given to the time of day. Into the 16th and 17th centuries, however, commercial activity was becoming increasingly important. Many small villages with growing populations were becoming centers of trade. In time, as industrialization became a force that pulled ever larger numbers of workers into factories, work schedules were established and the hourly wage became the common measuring unit for pay. With all of this change coming into secular life, more and more people were fixing their attention on the clock. Time would no longer be a consideration only for religious life. Time was becoming money.

Mechanical Clocks

Mechanical clocks are thought to have first appeared late in the 13th century or early in the 14th. They were initially powered by a falling weight connected by a rope wound around the axle of a toothed wheel (the "escape wheel"). The teeth on the wheel engaged alternately with upper and lower paddles (the "pallets") mounted on opposite sides of a vertical rod (the "verge"). The wheel's rate of rotation was regulated in the earliest mechanical clocks by an attached balance bar (the "foliot") on which regulating weights could be adjusted to change the rate of the rod's oscillation and, consequently, the wheel's rotation.

This crudest version of the mechanical clock worked but left much yet to be learned, both to allow more portability (the descending weight was a big hindrance) and to improve accuracy. The portability issue was confronted early in the 15th century when mainsprings began to appear as a replacement for the descending weight to drive the timing mechanism. Clocks with mainsprings could be moved while running. Not long thereafter, a balance wheel began to replace the rod, or verge, that controlled the rate of oscillation of the toothed wheel, and hence the rate of the clock. By the 16th century clocks small enough to be carried as watches came on the scene as yet another symbol of wealth. The problem with accuracy remained, even though a growing number of clockmakers, in competition among themselves, were striving to refine the craft.

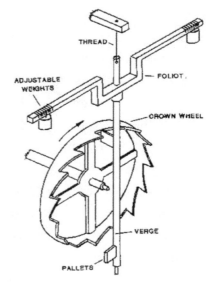

5.2 Schematic of mechanical clock
The falling weight to drive the clock was connected by rope to the lower axle holding the escape wheel. A mechanism to ring the clock's bells (not shown) could be affixed to the outer end of that axle.

The primitive mechanical clock was most commonly used to ring bells, both in church belfries for canonical purposes and in public towers. Its major flaw, however, was a lack of accuracy. The workings of the clock could only be refined enough to do an adequate job, not a suitably accurate one. Perhaps in part for that reason, only a very few of the early mechanical clocks, principally those most carefully constructed for astronomers, had a clock face that would allow a viewer to get a time reading. Bells would ring on the hour, more or less, but most of these clocks left the minutes and most definitely the seconds to a good guess.

Pendulum Clocks

Four hundred years! The pace of life and the need for good timekeeping were building. It was not until the 17th century, however, that the mechanical clock could be reconfigured to give it much improved accuracy. One reconfiguration introduced the pendulum to the inner workings. Pendulum clocks quickly replaced the earlier primitive devices, largely reliant for their accuracy upon the skills of the clockmakers, and became the mainstay for timekeeping.

Early in the 17th century, Galileo, the Italian astronomer best known for getting himself into trouble with the Catholic Church by putting the Sun rather than the Earth at the center of our solar system, investigated the motion of a pendulum. He discovered the key property that the duration of a pendulum's swing, one tick-tock, does not change with the arc of the swing, no other interferences being present (interferences begin to appear, however, as the arc increases). It is only the length of the pendulum that affects the duration of swing, and it is this property that makes it useful in a clock. The accuracy of the clock can be adjusted by changing the length of the pendulum, extending or shortening the tick-tock.

Galileo had the idea of a pendulum clock and his son partially constructed one, but neither lived to finish it. The actual invention of the pendulum clock came in 1656 in a design by Christian Huygens, a Dutch astronomer. Numerous improvements followed, and by the mid-18th century carefully-built pendulum clocks achieved accuracies of a few seconds per week.

Like the earlier mechanical clocks, a pendulum clock was most typically driven by a descending weight attached to a notched wheel (later, in some clocks a mainspring would be substituted for the weight). In the place of the older vibrating bar mechanism, a pendulum with its regularity of swing interacted with the notched wheel to regulate the clock. Through mechanical linkages, adjusting the length of the pendulum would change the speed of the clock. With the much greater accuracy this allowed, a second hand could be affixed directly to the notched wheel and, with suitable gearing, both a minute hand and an hour hand could be included to move in coordination with the second hand.

Balance Spring, No Pendulum

Alongside the advent of pendulum clocks came the separate concept of marrying a spring to the balance wheel. This was a solution for clocks that needed to be portable and, thus, could not use a free-swinging pendulum. When the balance wheel and the newly-conceived spring were skillfully matched, the two operated in unison as a natural oscillator, largely overcoming the earlier difficulty when the balance wheel alone was a slave to the gradual winding down, and thereby to the declining strength, of the mainspring.

Clock Face

As clocks gained accuracy, the clock face with dials became commonplace. The dial had been foreshadowed by the sundial, on which midday appeared at the top and, as the Sun moved westward, the passing hours were indicated by the shadow's movement eastward on the face. Just so, on even the earliest pendulum clocks noon was positioned at the "12", at the top of the face, and the passing hours, minutes and seconds were sequentially indicated in the direction which came to be called "clockwise."

Clocks Prevail

As lives became increasingly intertwined in a rapidly changing and ever more diverse economy, both pendulum clocks and portable balance wheel clocks became ever more prevalent, finding places in both dwellings and public gathering spots. Those who could afford them wanted clocks that would show their wealth. Countless styles of clocks, some with elaborate ornamentation and others with enhanced precision, came into being out of the growing number of clockmaker shops and factories. This predominance of both pendulum clocks, not portable, and balance wheel clocks, including watches, which were portable, continued for more than 250 years, up into the 1930s.

The tyranny of time that we are living under was dramatically and tragically demonstrated during World War I. Many British men had considered wrist watches to be "unmanly," but watches were issued to infantry soldiers as standard military equipment. The battle of the Somme began when hundreds of platoon leaders blew their whistles just as their synchronized watches hit 7:30 AM. The British suffered 60,000 casualties on the first day of that battle, most during the first hour. The tyranny of time has both good and bad consequences.

Clocks in China

Recall from the discussion about calendars that Jesuit missionaries in the late 16th century took the newly-adopted Gregorian calendar to China. Along with those calendars they took mechanical clocks, for presentation to Chinese rulers. These clocks, of course, would have had the older vibrating rod workings because neither the pendulum clock nor the balance wheel clock, with much greater accuracy, was yet to be invented. Even so, the early mechanical clocks from Europe were received with great delight in China.

Centuries earlier, up until the end of the 11th century, China had been in the forefront of building increasingly elaborate water clocks. The use of that device had been essentially astrological. A sighting tube mounted to the water clock made it possible to know the positions of celestial bodies when an emperor's wife or concubine produced an offspring. Five hundred years later, at the turn into the 17th century, the arriving missionaries found no trace of the water clock in China, even though that ancient timekeeping device was still not difficult to find in the western world, mostly as a symbol of wealth.

It is reasonable to assume that as years followed, with trade increasing between China and the western world, advancements in timekeeping and other technologies with growing importance in everyday life were finding their way back and forth between the two trading spheres. Timekeeping was increasingly important in commercial life at both ends, but by the same token it was increasingly important in everyday affairs.

Timekeeping at Sea

The pendulum clock brought a major improvement to timekeeping on land but it left mariners without a solution for their longstanding difficulty at sea. In short, the pendulum clock did not have sea legs. To work accurately it had to be kept stationary and level. Tossing about with the ship made it essentially unreliable.

The difficulty for mariners was in determining the longitude of the ship, the distance east or west of a known prime meridian. Latitude, the distance north or south of the equator, could be determined on board by making a sighting of the Sun or a star, using a cross-staff, an astrolabe or, beginning in the 18th century, a sextant.

Determining longitude, the east-west coordinate, was more difficult. It had to be measured against some point of known longitude on the globe.

The Cape of Good Hope, near the tip of Africa, was rounded in 1488 by a small fleet of three Portuguese ships, under the command of Bartholomeu Diaz, an explorer who had been dispatched to find a sea route to India. After Diaz reported that a route existed around the Cape, his countryman Vasco de Gama was dispatched in 1497 to make the first trade voyage. That opening of sea routes between Europe and Asia was a call for more east-west sea travel, and that pursuit put increasing importance on finding an onboard way to determine longitude.

The continuing hazard of sailing without a good reading of longitude was most poignantly driven home in 1707, more than 200 years after the de Gama voyage, when two thousand British seamen perished after all four of the ships in their fleet capsized after striking ledges on their voyage home from Lisbon.

To inspire talents to find a solution, the British Crown in 1714 offered a large monetary reward (today's equivalent of about $5 million) for inventing a method to know on board ship the time difference between the ship's location and the Royal Observatory in Greenwich. Jumping ahead to illustrate the method ultimately chosen, the ship's onboard (local) time would be determined by using astronomical instruments and nautical charts to fix the vessel's position relative to the Sun or other celestial bodies. The time at Greenwich would be determined by reading the sea-worthy clock (the chronometer) brought from port. If the ship's onboard (local) time were three hours behind Greenwich time, the ship's longitude would be 45 degrees west of Greenwich. The navigator on board would make that determination by reasoning, first, that three is one-eighth of 24 (total hours around the globe) and one-eighth of 360 (total degrees of circumference around the globe) would be 45 degrees and, second, the direction from Greenwich would be to the west (the ship's time was behind, not ahead, of Greenwich). To the extent navigational hazards at sea and coastal ports had been accurately charted, the vessel could be navigated to avoid the hazards and reach its destination port, following the safest route.

The standard for winning the Crown's reward was to be within 30 arc minutes in longitude (at the equator, roughly 35 miles east-west) at the end of a voyage from Britain to the West Indies. Several took up the Crown's challenge. The ultimate winner was John Harrison, a carpenter turned clock inventor. With support from the Royal Academy, his success came in the 1760s by making a watch, about five inches in diameter, that used a balance spring mechanism to control its speed. A ship's motion at sea had little effect on the accuracy. Harrison's device, called a chronometer, proved to be three times more

5.3 Harrison's first marine chronometer, weighing 72 pounds. It was completed in 1735 and tested at sea in 1736,. The much smaller instrument in the center foreground was Harrison's fourth effort (1761), which won the prize for determining longitude at sea. The other two chronometers at left and right were made by another watchmaker.

accurate than that required by the standard set by the Crown. When the full reward, however, was not forthcoming from the Board of Longitude, Harrison successfully petitioned the Crown to collect it.

Modeled after Harrison's watch, the marine chronometer had a great influence on navigation charts. By 1825 it had become standard equipment on all ships of the Royal Navy. Logic dictates that ships of other seafaring nations took equal advantage of Harrison's effort. Sea trade and travel had finally been put on course.

Now, in the 21st century, maritime navigators have available multiple systems for determining position and setting course. The marine chronometer and celestial sightings have continued to have their customary roles, while they have been joined by radio waves, radar, and satellite (especially GPS). Two or more of these systems are commonly deployed on a single voyage.

Timekeeping Marches On

Sundials used the daily rotation of the Earth around its polar axis. Water clocks used gravity's unchanging pull on water or another fluid. Pendulum clocks used the uniform swing of that pendant. Balance spring clocks used the back-and-forth of the balance wheel and its interconnected spring. In each of these progressions toward ever better accuracy, the regularity of a clock's action, which is to say its periodicity, the essential characteristic of any clock, was controlled either by a repeating pulse or, in the case of the water clock, a steady one-directional force that by mechanical connections could be made repetitive.

Quartz Clocks

Working from scientific discoveries in the fields of electricity and atomic structure that had come during the 19th century and the

> The generation of alternating electric charges within a crystal is called the piezoelectric effect. It has multiple uses unrelated to timekeeping, such as converting a voice into electric signals inside a microphone.

first decades of the 20th, scientists during the 1930s came upon using electricity to trigger oscillations within a crystal of quartz. With each oscillation, positive and negative electric charges would appear on the outer edges of the crystal. Once the frequency of the oscillations had been determined, the alternating electric charges on the edges could be captured by a second electric circuit which, through sequences of circuit boards, could then either drive a motor to turn the hands around the face of a clock or flash the time on a digital display.

While quartz clocks have become commonplace, some electric clocks, typically those that are least expensive, have no internal crystal to control their rate. They rely instead upon the oscillation of the alternating current (positive to negative to positive to negative, most typically at 60 cycles per second, or 60 hertz) in the electric grid that delivers power to the wall receptacles where the clocks are plugged.

As had been the case when the mainspring first appeared in the 15th century as a replacement for the descending weight in the earliest mechanical clocks, the discovery in the middle decades of the 20th century that the quartz crystal could be used to keep time brought with it an imperative to miniaturize the device for usage in a watch. With profits to be made, that miniaturization did not take long. It is a good bet that most of the wrist watches on Flight 1230 to Beijing were quartz crystal timekeepers, powered by coin-sized batteries.

Atomic Clocks

Never satisfied, scientists continued their efforts to improve the accuracy of timekeeping. And they have done just that, exponentially speaking. Work that started in the United States in 1945 and was then continued in Britain culminated during the mid-1960s in the construction of an atomic clock, with an accuracy better than one second per one million years. This timepiece harnesses the jumps of the

> Scientifically speaking, the atom in an atomic clock is caesium 133 at its ground state (lowest energy level). The oscillation occurs in its outer orbital ring, where a sole electron, when excited in a magnetic field, jumps between energy states. Those jumps occur at a resonant frequency of 9,192,631,770 cycles per second.

5.4 Caesium atomic clock, c. 1965

sole electron in the outer ring of a caesium atom. Those jumps occur with a stunning regularity of more than nine billion times per second. When this hard-to-imagine rate is captured and then slowed down by electronic circuits, the accuracy of the final output, in control of a connected clock, is about one million times better than what can be achieved with astronomical techniques.

For timekeeping, 1967 was a landmark year. Meeting in Paris that year, the CGPM, the governing body for BIPM, formally adopted a new definition of the "second", based upon the atomic clock, and made that newly defined unit the basic unit of time. Other units of time, such as the "minute" and the "hour", are now extrapolated from that revised definition of the "second." Salute the caesium atom.

> By its new definition, the "second" is "the duration of 9,192,631,770 periods of the radiation corresponding to the transition between the two hyperfine levels of the ground state of the caesium 133 atom."

As previously noted, scientists are never satisfied. Work is ongoing to harness for timekeeping the frequencies within other atoms where the electrons jump at a much higher rate than in a caesium atom. Eyes in laboratories are on accuracies that might be one second in five billion years or even one in 14 billion. We should stay tuned.

Giving "second" the new definition in 1967 was not a minor event, even if it was not the stuff of front-page headlines. It trumped all of the earlier centuries of efforts which had been devoted to extracting the best possible accuracy for timekeeping from astronomical observations. Astronomers had come to recognize that the Earth's rotation is gradually slowing down. The average duration of a day is becoming infinitesimally longer. More than that, the Earth wobbles due to oceanic tides, tectonic drift, and other forces. Astronomical measurements are perturbed by these irregularities now found in what for centuries had been taken to be regularities, the Earth's orbit and the Earth's rotation. Astronomy, it has been learned, has its own inherent limits of accuracy.

The newly defined "second" has transformed the determination of all units of time from the cosmic-scale, the Sun's motion relative to Earth, to the sub-atomic scale, the jumps of the outer electron of an atom. The greatly enhanced accuracy that this 1967 transformation made has already yielded technologies which serve inhabitants around the globe. The extreme accuracy of an atomic clock, of course, is not needed in wall clocks or wrist watches for the management of everyday life, but that accuracy is essential for many behind-the-scenes functions which do express themselves in everyday life, such as the determination of locations displayed on our GPS devices. Other critically important functions that rely upon highly accurate clocks are those of power grids, broadcast stations, mobile phone networks, and space exploration.

A new nomenclature has come with the transformation of timekeeping from the heavens to the atom. The new time standard is called Coordinated Universal Time. (The abbreviation is UTC. The French term for Coordinated Universal Time is *"temps universel coordonné"*. The abbreviation UTC represents a compromise between advocates for an English acronym, "CUT", and those for a French one, "TUC"). This new standard is actually a modified version of atomic clock time.

Leap Seconds

The BIPM in Paris maintains UTC, on a worldwide basis, by comparing readings from about 200 atomic clocks around the globe

(International Atomic Time, or TAI) with astronomical readings of solar time (now called Universal Time). Before atomic clock readings race ahead of astronomical readings by a full 0.9 seconds, the BIPM adds a "leap second" to UTC to keep it synchronized, more or less, with the Earth's gradually slowing rotation.

This ongoing effort to keep the new atomic time in synch with older solar time is an accommodation primarily for those who navigate the seas. When atomic time was introduced during the 1960s, sailors insisted that it be synchronized with the chronometers they were using to determine longitude. The solution was to add leap seconds occasionally to atomic time, which in effect notches back the readings of atomic time to keep it from gradually running farther and farther ahead of solar time. Without the leap second antidotes, over the course of a century the atomic time reading would become about a minute ahead of solar time readings.

> Scientists use UTC as needed, but most of us have little or no need. If we hear "4:00 Eastern Standard Time," we can simply pause to appreciate that time is accurately measured now by electrons flipping about inside an atom and our clocks are adjusted to account for small irregularities in the Earth's motions. The American military uses a 24-hour clock, calling it "zulu," which is UTC time geared to Greenwich.

Chronometers and, indeed, most other clocks now are set to follow UTC, so putting an additional leap second into their readings displaces an existing UTC with a new UTC (stepped back a second). Then the chronometers and clocks can continue to run at the rate of atomic time, but occasionally will be stepped back again by one second to stay more closely in synch with slowing solar time. These modifications of UTC occur roughly about every two years, as an insertion of an extra final tick at the end of the last day of June or December, on Greenwich time. The BIPM in Paris calls the shots. (Leap seconds, of course, could also be deducted to change UTC. This would occur if the Earth's rotation for some period were faster, not slower, than the atomic time. That has not yet occurred.)

Time Zones

Most probably as a vestige of the sundial, over the roughly five centuries that followed the introduction of the mechanical clock in the 13th century, the users of the newer timepieces paid allegiance only to local time. Noon was noon when the shadow cast on the face of a sundial, properly aligned for its longitude, pointed straight ahead (in the northern hemisphere, straight north). And long after the sundial had been largely replaced by the mechanical clock, it would remain true that noon was noon when both the hour and minute hands on the clock were straight up on 12 at the top of the clock face (symbolically, straight north).

Mechanical clocks and their miniatures, the watches, were regularly reset to the local time that commonly appeared on a public timepiece, most commonly in the town square. That timepiece, in turn, was regularly reset to an astronomical reading, perhaps one from a sundial or, in larger localities, an observatory. By centuries of habit, local time as displayed on public timepieces became engrained into everyday life.

Well into the 19th century, travel between towns was an arduous undertaking over rough and tumble dirt roads. For that reason and many others, personal travel was a relatively infrequent occurrence. Upon arrival at a destination, no matter the distance covered, the traveler would know to reset the clock in his case or the watch in his pocket to a new local time.

As can be easily imagined, the globe was a hodge-podge of local times. As countless localities were becoming increasingly interdependent, confusion and inconvenience were mounting. Over time the multiplicity of local times became a widespread problem that ultimately demanded attention, not that of clockmakers, for this problem was neither of their making nor within their reach, but of politicians and diplomats.

England

In 1784, England launched a nationwide mail-coach system, an over-the-road network that required coordination of timekeeping among its many local stations. Countrymen objected to having London time trumping their own local times, so those operating the new transportation system provided each of its coach drivers with timepieces that could be pre-set to the numerous local times along the route. Paradoxically, the new mail-coach system may have been another inspiration for the increasing number of English citizens who were moving towards larger cities. During the late 18th and early 19th centuries, industrialization was taking hold and the increasing integration of rural and urban life made it ever more difficult for smaller populations to stick with their own local times.

By the middle of the 19th century, in England a network of railroads was taking the place of the mail-coach system, and in 1840 a nationwide railway time was put into place. The time at Greenwich, the site of the Royal Observatory, was telegraphed in lines that followed the rail tracks and soon became the uniform "railway time" for the entire country. It was not until 1880, however, that Parliament stepped up by giving legal sanction to Greenwich time, making it the nationwide local time.

United States and Canada

Across the Atlantic, a similar history of timekeeping was being written in England's former colonies. With a much larger territory to be covered, numerous small railroad companies were springing into service. Each railroad typically fixed its hub in an urban center and used the local time of that center to prepare and publish its own operating schedule. A resident of a town along the route that spoked out from a railroad's hub would need to know the time difference between his locality and the hub city to be sure he would get to his local station on time to catch a train. Countrywide, railroads were using over 80 different time standards.

Inconvenience and confusion were one problem, but safety was quite another. The safety record for trains in the United States during the 1830s and 1840s was not perfect, but not overly alarming either. That changed with a sequence of collisions that occurred in 1853 and following years. In most cases the collisions could be attributed to reliance upon inaccurate timepieces in the hands of conductors and train engineers. In some cases, the lack of safety margins in tightly scheduled operating schedules shared the blame. Certainly, wherever tracks were shared by two or more railroads, each operating on its own time standard, the likelihood of human error was compounded.

The railroad industry put its arms around the problem. In November 1883, United States and Canadian railroads jointly converted their operations to "Standard Railway Time," a system that divided both countries, east to west, into four time zones with a one-hour separation from zone to zone. This adoption of four zones anticipated the eventual concurrence of most nations with the concept of 24 time zones to encircle the globe.

Over the years that followed, many localities in North America abandoned allegiance to their own local times and put themselves in synch with Standard Railway Time. In some spots, however, especially those near the dividing lines between time zones, where the switch to Standard Railway Time brought a more drastic change to the clocks, resistance to the change continued. State legislatures began to adopt Standard Railway Time for their own governmental purposes but stopped short of dictating conforming change for localities. Eventually, however, the switch to Standard Railway Time became widespread, even in those localities where resistance had been most entrenched.

Europe

Returning to Europe, railroads which had earlier begun service on the continent using a patchwork of local times followed the examples which had been set by their counterparts, first in England and then North America. But going beyond the need for a uniform time system for rail operations, a highly-respected Prussian field marshal, in an 1891 address to the German Reichstag, further argued the need for uniformity to manage and mobilize the military. In a short period of time, unification took hold across much of Europe, with time zones one hour apart and Greenwich as the prime meridian. The practice of local time faded into history.

France was a laggard. Reflecting its national pride, in 1891, when its neighbors were changing to a uniform time system using Greenwich as the prime meridian, France, both within its national borders and in colonized Algeria, was holding to its own legal time using Paris as its reference meridian. It was not until 1911, about 20 years and much dithering later, that France came into line.

Clocks

China

Until the early 20th century, timekeeping across China held to local times determined by astronomers in the capitals of the imperial dynasties. Geographically, the country spreads across what would normally be five time zones. In 1918 the central observatory in Peking (now Beijing) proposed dividing the width of China into five time zones, the easternmost of which would have a time difference of 30 minutes, not a full hour, from its neighboring zone immediately to the west and the westernmost similarly would have a time difference of 30 minutes from its neighboring zone to the east. The three intermediate zones would each be separated from one another by a one hour difference, like the rest of the globe. That division was adopted in 1939, more than 20 years later, but was abolished in 1949 after the Communist Party had taken control.

Apparently for the purpose of unifying what had historically been a population with multiple ethnicities, using different languages and ruled by different dynasties, the Communist regime pulled all of China into a single time zone, that of Beijing. (Domestically, and especially within the majority Han ethnicity, this is called Beijing Time, while internationally it is called China Standard Time.) In its farthest west region, official Beijing time puts the clock two hours ahead of what local residents would otherwise deem to be the local time. Meeting a friend for lunch at "noon" can mean meeting when it is 12:00 noon in Beijing (10:00 AM local time in the west) or when it is 2:00 PM in Beijing (12:00 noon local time). Someone's lunch might get cold!

At Sea

At the invitation of the United States, 25 countries sent their representatives to the International Meridian Conference held in

Washington D.C. in 1884. The goal of the conference was to give endorsement to Greenwich as the prime meridian for determining longitude around the globe and as the reference point for a universal time. That goal was achieved, with 22 of the participating countries approving Greenwich, but one country disapproved and two others, including France, abstained. France had argued unsuccessfully for Paris to be the prime meridian. It came around to official acceptance of Greenwich in 1911, but it did so in French style, not by using the word "Greenwich" but instead adopting "Paris Mean Time, retarded by 9 minutes, 21 seconds."

The approval given to the Greenwich meridian at the 1884 conference was consistent with what had already become the prevailing practice among mariners of all nations, in large part because most were using navigation charts published by the British Admiralty. Implicitly but not expressly, putting the prime meridian at Greenwich also located the anti-meridian, the international date line, at the 180 degree meridian up and down the Pacific. Although urged by some to do so, the conference did not undertake to define worldwide time zones.

The action taken in 1884 by the International Meridian Conference was not binding, even for the nations that had participated. The legislative bodies in Britain and Japan took prompt steps to adopt Greenwich as the prime meridian, but their counterparts elsewhere, including the United States Congress, were reluctant to follow. The long-argued public dispute about the prime meridian went on, fueled by proponents of various alternatives to Greenwich. For all practical purposes, including navigation at sea, however, Greenwich had been chosen.

If railroads gave good cause for time uniformity on land, the advent of wireless radio transmission during the 1890s gave good reason for uniformity at sea. Starting in May 1907, signals generated by an observatory in the Canadian province of New Brunswick were automatically transmitted from Halifax Harbor in Nova Scotia. By 1910 high-powered stations on the North Sea in Germany and the Eiffel Tower in Paris were transmitting to broad swaths of the Atlantic as well as much of Europe.

In the years that followed, automatic transmissions to ships at sea of observatory times on shore became a common navigational aid worldwide. These radio signals of observatory times were used on board ship to check the accuracy of the chronometers from which the ship's longitude could be determined. Subsequent decades have brought several more advances to marine navigation, not the least of which is the global positioning system, the operation of which is critically dependent upon maintaining precise timing among its multiple satellites.

Daylight Saving Time

Daylight saving time (commonly signaled by "DST") puts kinks in the push for a single timekeeping system around the globe. About 140 countries have used daylight saving time (often with their own nomenclature, such as "summer time") at least once, but only about 75 are still doing so. Even among those that do, the starting and ending dates may well differ. More than 100 countries have never used it. In Europe, where daylight saving time has been followed since the 1980s, the European Commission in 2018, reacting to a survey, signaled that its member nations might be allowed to choose their own time zones. A large majority of respondents favor scrapping the twice-annual time changes, due in part to health problems thought to be associated with them.

Countries in the southern hemisphere experience their summers, of course, during the winter months of the northern hemisphere. Logically, those who live below the equator in a country that uses daylight saving time need to adjust their clocks, and their internal senses, at times opposite of those who live north of the equator. In China, daylight saving time was used in a few localities earlier in the 20th century and then nationwide from 1986 to 1991, but it has had no role since then. Countrywide, China is all one time, one zone, no daylight saving time.

Like the earlier movement to replace multiple local times with a single time countrywide, the notion of daylight saving time took root in England. The idea came from William Willett, a homebuilder and early morning equestrian, noticed as he rode by after daybreak that window blinds were closed on homes to block out the morning sun. An essay he published in 1907, titled "The Waste of Sunlight," led to consideration in Parliament of a proposal to move the clock ahead during the months with days of longest sunlight. Three groups opposed: farmers and dairymen, newspapers, and public workers' unions. Willett also made his pitch to members of Congress in the United States. No action was taken by either Parliament or Congress.

World War I brought attention back to the concept. In April 1916, Germany and Austria-Hungary advanced their clocks by one hour, ostensibly to reduce the consumption of oil and coal. Over the next two months similar changes to advance the clock were made by several Allied nations, including Britain. After the end of the war, Britain continued to use what was called "Summer Time" throughout the 1920s and 1930s, even though the original justification of conserving fuel for wartime had dissipated. During World War II, Britain doubled down by putting its clocks two hours ahead to "Double Summer Time" during the summer and one hour ahead during the winter. After the end of that war, Britain dropped the doubling and reverted to "Summer Time." During its EU membership, the nation's beginning and ending dates for DST have been set with EU mandates. With Britain now expecting to exit the EU, the question arises whether it will change its dates of beginning and ending DST.

United States

In the United States, daylight saving time in its early years had an on-and-off, spotty experience. It first captured interest in localities where the change to Standard Railway Time had been most disruptive. In 1909, Cincinnati, situated on the eastern edge of what was then the Central Time Zone (as demarcated in Standard Railway Time), by action of its city council, adopted daylight saving time for May through September, beginning in 1910. That local action, however, was repealed in April 1910, only two weeks before it was to become effective. Similar proposals to adopt daylight saving time came forth in subsequent years in New York and other localities, and other proposals were made in Congress, but none was enacted.

As had been the case in Europe, daylight saving time on a national scale came to the United States in March 1918 by action of Congress, over resistance voiced on behalf of a railroad industry which did not want to alter train schedules twice a year. The country had entered World War I in April 1917. The arguments made to Congress by proponents of daylight saving time had been both the wartime need for fuel conservation, an argument which had been successful in Europe, and the support of food production by extending daylight for evening gardening.

This early flirtation with nationwide daylight saving time in the United States had a short life. Over President Wilson's veto, Congress in 1919 repealed its March 1918 wartime action. The issue then moved to state legislatures and local governments. From 1919 onward, except for a period during World War II, the citizens of any particular locale across the United States were vulnerable to decisions in a federal regulatory proceeding (redraw a standard time zone), or enactments by a state legislature (adopt, change or repeal daylight saving time), or actions by their local governing body (adopt, change or repeal daylight saving time). Any of these steps might have required the affected citizens to adjust their clocks and, consequently, their internal sense of the time of day. More than that, they had to be aware of time changes that might have occurred at other localities where they had contacts, whether commercial, family or social. A common question: "It's 5:15 here, what time is it there?"

During World War II, Congress took action a short time after the bombing of Pearl Harbor to place the entire country on year-round daylight savings time, advancing all clocks by one hour. From February 1942 until September 1945, the four time zones across the nation were called Eastern, Central, Mountain and Pacific "War Time." As has been mentioned, during the war Britain went for a two hour nationwide advance, called "Double Summer Time."

Except for that wartime action, Congress kept its hands off the issue of daylight saving time for almost 50 years. In 1966, however, it stepped up by mandating nationwide observance of a one hour advance from the last Sunday in April to the last Sunday in October. Individual states, however, were given the right to opt out, either for the entire state or, if the state straddled two time zones, for the entire portion within one time zone. Arizona and Hawaii have done just that for their entire states (in Arizona, however, the Navajo Nation in the northeastern corner observes daylight saving time). The federal regulatory process for altering time zone boundaries when good cause is shown now lies in the Department of Transportation. The process has been invoked several times on petition by individual states, most often with the result of edging time zone boundaries westward.

Since 1966, Congress from time to time has changed the starting and ending dates for daylight saving time, the most dramatic of which were the extensions to 10 months in 1974 and then to eight months in 1975, both made as a fuel saving effort in reaction to the 1973 oil embargo. Each change by Congress has come with controversy, the primary argument of proponents being to energy and that of opponents being to avoid sending children off to school into early morning conserve traffic before daylight.

Worldwide Usage, With Exceptions

Timekeeping as we now know it started long ago in self-dependent localities which grew their own crops for self-survival. Over the course of centuries, that self-dependence gave way to increasing trade and transportation, and clocks, initially crude, became increasingly important. It was not until the 19th century, however, primarily under pressure from the railroads, that the practice of each locality using its own "local time" was replaced in favor of all communities setting their clocks according to a nationwide, and soon thereafter a worldwide, method of timekeeping.

But the concept of worldwide uniformity in timekeeping still has major blips. The largest is in China, where since 1949, when the Communist Party came to power, that nation has maintained a single time zone across its east-west expanse that geographically would otherwise encompass five time zones. The declared purpose of this was and is to unify diverse cultures residing under a single government. Other blips are found along the 180° line of longitude in the

Pacific Ocean, where inhabitants of some island nations see themselves to be more closely aligned with cultures on the other side of that line, and their clocks are set accordingly. Within the United States, federal law leaves room for individual states to disregard daylight saving time and two states, Arizona (excluding the Navajo Nation) and Hawaii, have done just that. The end result, globally, is that a uniform timekeeping system has generally been accepted, but exceptions exist where a government with the power to do so has decided to carve out an exception for itself.

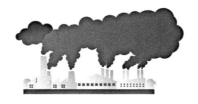

Chapter Six

TEMPERATURE

When Flight 1230 took off from Washington-Dulles, the ground temperature was **35°F,** or **2°C**. The flight crew used this reading, together with total aircraft weight, wind speed, wind direction, Washington-Dulles field altitude and other factors, to calculate the speed at which the plane could safely roll into its lift from the runway.

About 30 minutes into the flight, the pilot came on the speakers to greet the passengers and give a short overview of the flight plan. Among other points, he mentioned that Flight 1230 had a planned maximum cruising altitude of 39,000 feet, or about 12,000 meters, and at that altitude the outside air temperature would be approximately **−70°F,** or **−56°C**. The pilot then reported that the weather at destination in Beijing was forecasted at arrival to be scattered clouds with a temperature of **4°C,** or **39°F**.

Inside the cabin, the passengers were being served snacks and beverages while settling into their seats for the long flight ahead. The cabin temperature was **72°F,** or **22°C**.

Temperature Measures Heat

Curiously, like the pilot of Flight 1230, we talk about "temperature," not "heat." Temperature is a measurement of heat, and our real interest is in heat. If we accidently touch a hot pan, we are burned by the heat in the pan, not its temperature. If a hiker becomes lost and dies from hypothermia on a frigid mountain trail, the death is due to the loss of body heat, not the low temperature on the slope. We talk about temperature because we have instruments that can measure it. We do not have instruments that can measure heat, per se, so we accept temperature as an expression of heat. Cold, of course, is what we call a shortage of heat. We measure what we can, and we talk about what we can measure.

Measuring heat is the function of thermometers. (The English word "thermometer" comes from the French *thermomètre*, which was coined in 1630 by a Jesuit priest using the Greek words *thermos* (hot) and *metro* (measure).) Scientists and inventors have provided us a range of devices to detect, quantify and display temperature. Many thermometers display their measurements both in degrees Fahrenheit (°F) and degrees Celsius (°C), while others display them only on one of these scales, not both. We have been taught how to make the conversions:

°F = (°C x 9/5) + 32 or going the other way °C = (°F − 32) x 5/9

Although not often encountered in everyday life, it deserves mention here that scientists use a third scale of temperature which expresses the measurement of heat simply as kelvin (symbol K), without the word "degree." As many already know, the names of our three temperature scales, Fahrenheit, Celsius, and kelvin, come from the three men in the history of thermometry who are credited with proposing them. More later.

First, we should understand that *heat* is the underlying phenomenon that makes thermometers so important, not only in our everyday lives but more so in science and industry. Heat is a fundamental form of energy. Most simply described, it is the level of commotion among the individual, microscopically tiny atoms and molecules (collectively, the particles) that make up stuff we sense to be hot, cold or maybe just right. The porridge that Goldilocks tasted in the house in the woods was a stew of many particles (the stuff of meat and vegetables). Those particles were bouncing around in the three bowls, but Goldilocks could not see them because they were much, much too small. In the first bowl that was "too hot," the particles were bouncing against each other and the sides of the bowl with too much energy. In the second bowl that was "too cold," they were bouncing with too little energy. In the third bowl that was "just right," they were bouncing with just the right amount of energy. The thermometer that Goldilocks was using was her mouth. Crude, yes, but still, a thermometer of sorts.

Like the porridge, the air that surrounds us every day is too hot, too cold or just right, depending upon the amount of energy it contains, which is to say the level of commotion among its particles. When we step into the fresh morning air, we feel its heat. If we glance at the thermometer on the porch, which is "feeling" the same air, it will give us a reading, say 35°F, or 2°C, along with an appreciation for the jacket we are wearing. The jacket, like other clothing, helps us retain the heat our bodies are continuously generating (the billions of particles inside us are in perpetual commotion). The thermometer on the porch "feels" the temperature because heat can transfer itself between the ambient air and the liquid inside the thermometer, most likely tinted alcohol. Nature seeks to keep things balanced. If a thermometer is relocated from inside the warm house to an exterior wall on the porch, some of the heat inside the thermometer's liquid will escape into the porch air. As the liquid cools, it will shrink down the stem and the scale of the thermometer. That transfer of heat will continue until the thermometer's liquid and the porch air reach the same temperature, until they are "thermally balanced." A reverse transfer of heat would occur, of course, in the summer months if the thermometer were taken from the air-conditioned house into the hot afternoon air on the porch. As the liquid in the thermometer warmed, it would expand up the stem, showing a higher temperature on the scale.

    We are assuming that the thermometer on the porch is a common household type that relies upon liquid in an enclosed glass bulb and stem to feel and display temperature. The transfer of heat between the ambient air and the thermometer's liquid occurs by a process called conduction. Heat can also be transferred by convection or radiation. Put an empty pan under the kitchen faucet and then turn on the hot water. The process of transferring heat from the water heater to the pan is called convection (warm water carries the heat through the house pipes). Put a couple of steaks under the oven's broiler, or on the charcoal grill outside. The process that transfers heat from the hot broiler coil or the glowing charcoal to the steaks is called radiation (heat emitting from the broiler or the hot charcoal lands on the steaks). These three methods of transferring heat, conduction, convection and radiation, are all used to harness thermometers. The situation determines the method. The tendency of heat to balance itself out is the characteristic that makes it possible to measure heat with a thermometer. It is also the reason, obviously, that insulation is installed when a house is constructed. Nature's search for balance is sometimes helpful, sometimes not.

Temperature

Air pressure will affect the commotion among the particles in the air that strike the thermometer on the porch, and thus upon the temperature the thermometer will display. Higher air pressure on the porch means there is more commotion, which is to say more heat, in the ambient air. That yields a somewhat higher temperature. As should be expected, lower air pressure means less commotion, less heat, and lower temperature. In the morning air, this effect of ambient air pressure is not likely to make any noticeable difference in the thermometer's measurement.

> Wind chill is not the same thing as either air pressure or ambient temperature. It is simply an effect we feel on our exposed skin due to an increase in the loss of body heat. Moisture on the skin evaporates, and evaporation uses up body heat by continuously requiring the skin to re-warm itself. Thermometers ignore wind chill.

In a different setting, however, air pressure can be a much more significant factor. As Flight 1230 climbed to its cruising altitude on its flight to Beijing, outside air pressure declined because the air was becoming thinner at higher altitude (a lower density of particles banging around in the outside air means less commotion). The outside temperature would decline to −70°F, or −56°C, at 39,000 feet. The thinner air reduced the plane's rate of fuel consumption because it imposed less resistance to the plane's forward thrust. The fuel saving attributable to the thinner air aloft was offset to some degree, of course, by the increased use of fuel to keep the inside cabin comfortably warm while the outside temperature was falling to well below freezing. Even so, the net result for Flight 1230 at higher altitude was a significant reduction in the rate of its fuel consumption.

The point to keep in mind is that a thermometer, regardless of its type or the situation, gives us a measurement of the commotion among the microscopic particles within the stuff the thermometer is "feeling," be that the morning air on the porch, the hot water in the pan, or the steaks on the grill. We cannot see heat. What we can see is a thermometer's display of the heat it is feeling. That is our proxy for heat. We call it temperature.

Thermometers – In Everyday Life

Various types of thermometers inform our everyday lives. The weather thermometer on the porch, or the thermometer used by a weather reporter, helps us choose our clothing. The thermometer inside the thermostat on the wall tells the furnace or the air conditioner to heat or cool the household. The thermometer that monitors the automobile engine and sends its readings to the dashboard tells us if the vehicle is operating as it should. In the kitchen, both the refrigerator and the oven have thermostats, which are thermometers coupled to regulators wired to switches. The thermostat in the refrigerator holds the milk at its safe storage temperature. The thermostat in the oven holds the pot roast at the cooking temperature set on the control panel.

The boiling temperature of water will vary a few degrees depending upon the air pressure in the kitchen. That is so because the particles at the surface of water (primarily H_2O molecules) must have enough commotion to spring away from their fellow molecules in the water. If the air pressure above the water surface is high, particles need more commotion, which is to say higher temperature, to spring away and become steam (water vapor). If less, they need less. That is why an egg poaches a bit more quickly in Miami (lower altitude means higher air pressure, that means higher boiling temperature, and that means faster cooking) and why the same egg poaches a bit more slowly in Denver (higher altitude, lower air pressure, lower boiling temperature, slower cooking). Cooks learn this.

Thermometers – Outside Everyday Life

Temperature has an important role, perhaps an even more important one, outside everyday life. Science revolves around temperature. Much of commerce, industry and military is heavily dependent upon the capacity to measure and regulate temperature. The degree of accuracy required from the thermometers used in these many different realms varies with their purposes, but our central point remains true for all. Every thermometer, however designed, is used to measure the commotion among the particles, which again is to say the heat, that make up every gas, liquid and solid.

Science takes its guidance from the ongoing efforts at the BIPM, and at the several national laboratories that collaborate with the BIPM, to refine the measurement of temperature. In 1954, the BIPM made temperature, or more accurately the "kelvin", a measurement of temperature, one of its base units. This happened about 75 years after the BIPM had been launched with its focus initially limited to the meter and the kilogram. The kelvin, consequently, is now one of the seven units of measurement that can carry the designation SI.

Why the name "kelvin" (the official symbol for kelvin is K)? As will be explained in more detail later, kelvin is a scale for measuring temperature which was advanced in the mid-19th century by an English scientist-engineer, William Thomson, later to become Lord Kelvin. Why not "degree kelvin", just as we say "degree Fahrenheit" and "degree Celsius"? That is a choice made in 1967 by the CGPM, the international body that oversees the BIPM. Science, for its own reasons, often uses its own language.

The BIPM's definition, both old and new, implicitly puts zero K, shorthand for "zero kelvin", at absolute zero, which would be zero heat. That is as cold as it can get! But zero K is only a theoretical temperature. Scientists have come really close to zero K, but sucking away from any substance the very final bit of commotion among particles, the very last bit of heat, is likely always to be a fool's pursuit.

In the United States, the NIST collaborates with the BIPM by providing its "customers" (for the most part, other scientific organizations) with calibrations and reference standards. It also provides to the BIPM and the NIST's counterparts in other nations the results of ongoing efforts by the NIST's staff to improve the international standards for measuring temperature. The NPL in Britain, the NML in China and other national laboratories are doing the same in their own countries.

Backstory

The history of numbers and that of measurements of weight, length and volume take us back to ancient times. As seen in other chapters, that long retrospective is also true for calendars and clocks. It is not so much the case, however, for the history of temperature.

Philo of Byzantium, a Greek engineer and writer of the 3rd century BCE who spent most of his life in Alexandria, published a description of an experiment in which a hollow sphere (presumably sealed) held one end of a tube and a jug of water held the other end. If the sphere were placed in the sun, the warming air it held generated bubbles in the jug (the air was expanding). Conversely, if the sphere were moved to the shade, the cooling air caused some of the water in the jug to rise into the connecting tube (the air was contracting). Philo's experiment, however, apparently involved no effort to measure the temperature of the sphere. Three centuries later, in about

50 CE, Hero of Alexandria, a Greek mathematician who also spent much of his life in Alexandria, is thought to have modeled a "thermometer" on Philo's work. But neither Philo nor Hero dealt with what should be called a "thermometer."

Galen, a surgeon who became the physician of the Roman emperor Marcus Aurelius but is most widely known as the author of the med-

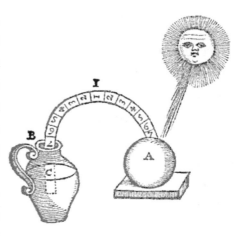

6.1 Philo's "thermometer"

ical book that prevailed in western medicine for about fifteen centuries, in about 175 CE devised an instrument which gave a crude measure of body temperature on a nine-point scale. This suggests, of course, that Galen saw some relationship between body temperature and health. After Galen, the recorded history about temperature seems to have gone blank for more than a thousand years, until late in the 16th century.

The Age of Enlightenment in Europe, roughly the 17th and 18th centuries, brought new interest and activity back to the subject of temperature, as it did to a broad sweep of interests. In retrospect, many who we might now call scientists of their day, plus many who maybe we should instead think of as inventors, gave a great deal of attention to the development of thermometers. They were to be found in several different places throughout Europe and, in many cases, they were working without knowing what others might have already learned.

Thermometers, Liquid-In-Glass

The attention of these pioneers was focused on the earliest type of thermometers, the liquid-in-glass type still commonly found in our homes. Other types would have to await the scientific progress that was yet to come in later centuries, especially the progress in how various materials other than liquids behave when they are exposed to changes in temperature.

There are only indirect indications in available sources about the motivations that played on scientists and inventors during the Enlightenment and later who put their focus on measuring temperature. Certainly, some, like Galen centuries before them, were motivated by an interest in medicine. They apparently sensed that measuring body temperature would be a useful diagnostic tool. Others, presumably, were motivated by a curiosity about weather. An ability to track temperature could have importance to farmers, travelers, and the population in general. And then, as should be expected, some were no doubt motivated by the very human instinct to distinguish themselves by being the first to make a scientific discovery, what perhaps too cynically might be labeled one-upmanship. One clue that one-upmanship had some role is the colossal size that some of the earliest thermometers were given. A height of almost 12 feet likely had no scientific cause, yet at least a few of that size most likely brought attention to their makers. More typically, much of the early development efforts can be assumed to have been centered on devices that would sit on a bench top.

Whatever the motivations might have been for the early explorers of thermometry, gratitude can be felt for the many efforts they made. Everyday life around the globe has been much improved, directly and indirectly, by the accumulation of know-how we now have, not only to measure temperature (think heat) but also to forecast it.

None of those who were to take part during the 17th and 18th centuries in the development of liquid-in-glass thermometers could have foreseen the critical issues to be encountered. As we know in hindsight, their collection of independent efforts would ultimately bring insights on three questions: first, should the thermometer be sealed or left open to the surrounding air; second, what would be the best fluid to put in the bulb and stem; and third, what would be the best scale for measuring temperature.

Galileo (1554 – 1642), the Italian who is best remembered for shaking Catholic doctrine by suggesting that the Sun, not the Earth, was at the center of the solar system, had a role in renewing scientific interest in measuring temperature. Some have credited him with the invention of the thermometer. Galileo's attention, however, may have gone no further than what he attached to a liquid-in-glass instrument that historians now refer to as the thermoscope. That was a vertically positioned glass tube with a sealed bulb at the top and an open end at its base, submerged in an unsealed jar of water. It could show if the temperature of the surrounding air had risen (the trapped air inside the bulb and stem had expanded) or fallen (the trapped air had contracted). The thermoscope, without a scale, however, gave no reading of temperature. Moreover, the unsealed jug at its base would have caused the expansion or contraction of the air to be affected not only by temperature but also by air pressure working on the water in the base. This instrument was a start, but much was yet to be learned.

It was not too far into the 17th century when a numbered scale was placed alongside the stem of the instrument, thus upgrading a thermoscope to make it a thermometer. The first known drawing of a thermometer is dated 1611. It shows a scale of eight degrees, with each degree further divided into "minutes" numbered 10, 20, and so forth up to 60. (Recall that the Babylonians, three thousand years earlier, had used numbers with the base 60, and then centuries later the faces of clocks were divided into 60 minutes per hour, 60 seconds per minute. Habits stick.) Interestingly, the highest temperature on this supposedly first thermometer was at zero. It took more than a hundred years to turn this around, so that hot would be the higher number and cold the lower. As in its predecessor, the thermoscope, the liquid in this early thermometer was water.

So the advancement of thermometry was underway, with multiple participants across Europe making their own efforts. Some of those efforts have no doubt been lost in the dustbin of history. A few names have become everyday words. It is important to keep in mind that the achievements of these few participants are only the headlines of a larger story.

Ferdinand II (1610 – 1670), a member of the Medici family who ruled as the Grand Duke of Tuscany in the mid-17th century, is credited with sealing the liquid in a thermometer to avoid the inaccuracy caused by the pressure of ambient air. Tuscany was enduring economic turmoil during his reign, but with a strong interest in science Ferdinand II used his position to attract scientists to Florence, the center of Tuscany's government, where much attention was given to improving the measurement of temperature. It is reasonable to suspect that credit for a discovery at least sometimes flowed upstream to the Grand Duke, even if he had not been personally involved.

By the middle of the 17th century, it had been determined that water would not be the best liquid to use in a thermometer. One characteristic of water is that it does not expand and contract proportionally with changes in temperature. No substance expands and contracts with changes in temperature along a straight line (linearly), but water has a particularly troublesome quirk. As temperature falls, water contracts until it reaches about 4°C (39°F), and as temperature then continues to fall water expands until it turns to ice at 0°C (32°F). (This explains why both icebergs and ice cubes float. They are less dense than the cold water that keeps them afloat.) Moreover, water becomes ice, a solid, at 0°C (32°F). We want most thermometers to keep working as temperatures go below that freezing point.

Two substitutes for water received most of the attention from early scientists and inventors. Diluted alcohol (what was called "spirit of wine"), typically tinted to make it more visible, was advocated by some because its expansion and contraction with a change in temperature (what science students are taught to call its coefficient of expansion) was greater than that of the second choice, mercury (called "quicksilver"). Mercury was favored by others because its response time to a change in temperature was more rapid than that of alcohol. Both types of liquids survived and became common in modern thermometers. As now commonly known, mercury is a toxin. For that reason, it has been largely eliminated from usage for medical purposes and has come under both federal and state regulations. That red line we see in our modern household thermometers, the liquid-in-glass type, is most likely tinted alcohol.

> Other liquids such as olive oil and saturated salt solutions drew the attention of early experimenters, but none of these alternatives gained a following.

The issue that brought the most diversity into the development of thermometers was the choice of the scale that would put numbers on the rise and fall of temperature. This actually had two sub-issues: first, what temperature points in nature should be used as fixed points to calibrate a thermometer; and second, once the fixed points had been chosen, how many divisions between those fixed points should be established to set the size of a single degree of temperature.

The advantage of uniformity, the use of the same scale of measurement, was recognized at an early stage. Some saw, as early as the beginning of the 18th century, that it would be useful, in the study of weather patterns, to have comparable temperature readings from different locations. But uniformity was not achieved. Working against it were the multiplicity of temperature pioneers, the paucity of communication links among them, and the instinctive human urge to own the final result. By the middle of the 18th century, thermometers were being made with up to twelve or more scales mounted alongside each other on a board behind their stems. In the 21st century, we are down to three scales in common usage: Fahrenheit, Celsius and kelvin. That is not uniformity, but at least we can readily make conversions among these three.

Fahrenheit Scale

Daniel Gabriel Fahrenheit (1686 – 1736), was born in Danzig (now Gdansk in Poland) but spent most of his life in Holland, where he became a well-known instrument maker. In 1708, while still at a young age, Fahrenheit visited an astronomer in Denmark, Ole Rømer, who had been experimenting with thermometers and who inspired his young visitor to try his own hand at their manufacture. Rømer had used alcohol in his instruments and, as an astronomer, had based his

6.2 Daniel Gabriel Fahrenheit

scale of temperature, which he modified from time to time, across the range from zero (the coldest that Rømer could imagine) to 60 (the boiling point of water). (As an astronomer, Rømer would have instinctively chosen the number 60 for his scale, that being consistent with 360 degrees around the globe, 60 minutes in an hour, and 60 seconds in a minute.) As the lower fixing point for his scale, he used the freezing point of water, which he set initially at 7½ and later raised to 8. By setting the freezing point above zero, Rømer intended to leave space on his scale, from 8 down to zero, for the below-freezing temperatures occasionally encountered in Copenhagen's winters.

In the years following his visit with his mentor in Denmark, Fahrenheit substituted mercury for alcohol in many but not all of the thermometers he was making. It is thought that Fahrenheit never understood that mercury and alcohol do not expand and contract in unison when measuring the same change in heat (neither of the two liquids expands or contracts along a straight line, and between any two temperatures they do not necessarily track together). He produced thermometers to

> While Fahrenheit was working in Holland, a well-born and self-assured scientist in France, René-Antoine Ferchault de Réaumur (1683 – 1757), was developing a thermometer using a scale of 80. His numerous publications and the respect he held within France and the scientific community across much of Europe meant that for more than a century his thermometers would be dominant not only within his own country but elsewhere on the continent, excluding Britain and Fahrenheit's home turf in the Low Countries. Réaumur's thermometers, however, did not make the final cut.

be used by his customers in science, medicine and the recording of weather, willing as a good merchant should be to satisfy their individual preferences for particular scales of measurement. As for his own preferences, Fahrenheit divided each of his mentor's 60 degrees into four parts. Having done that, the lower fixed point on his scale became 32 (4 times Rømer's 8). His upper fixed point was 96. Initially, Fahrenheit had chosen 90 but he later upped that to 96, what he thought to be normal body temperature. Fahrenheit's reasons for setting his own upper fixed point to be body heat instead of the boiling point of water are unclear. It is reasonable to assume that his choice was based upon years of experimentation, motivated in part by a desire to improve the instruments he was constructing for customers. In 1724, Fahrenheit put the boiling point of water at 212 on his scale. Following his death in 1736, both the boiling point and the upper fixing point for the Fahrenheit scale became commonly accepted to be 212. That has stuck for almost three hundred years.

Fahrenheit deserved the recognition he received not only for his work in thermometry but also that in meteorology and general metrology. Nevertheless, stepping back, the temperature scale that carries his name can fairly be called bizarre. It appears that Fahrenheit never really released his thinking about the scale from the influence of his early mentor in Copenhagen. His namesake scale obtained wide acceptance across Europe until well into the 20th century, by which time most nations around the globe had shifted to the Celsius scale.

In the 21st century, the Fahrenheit scale still has a firm hold on everyday thinking only in the United States, its territories and smaller countries which host United States military bases or are commonly visited by American tourists. The rest of the world's population thinks about the Celsius scale, with decreasing attention to Fahrenheit.

Celsius Scale

Anders Celsius (1701 – 1744) was a Swedish astronomer who became a professor at Uppsala University, north of Stockholm. After returning to Uppsala from an arduous expedition with other scientists to the far north of Sweden, in his mid-thirties Celsius had an increased awareness of the need for better thermometers to improve meteorological readings. By 1741 he had constructed his first thermometer. Befitting his scientific background, he used a scale of 100. (Celsius was thinking as a scientist. Sweden did not adopt the metric system for general use until 1889.) Interestingly, he used zero to denote the boiling point of water, not its freezing point, and 100 to mark the freezing point. By today's reckoning, Celsius had the scale upside down (as it had been on the earliest thermometers in the 17th century).

6.3 Anders Celsius

Apparently, upside down was the thinking of several of Celsius' contemporaries. After his death in 1744, his successor at the University, Martin Strömer, and a Stockholm instrument maker, Daniel Ekström, reversed his scale, making zero the freezing point and 100 the boiling point. But at least two more names belong here. Carl von Linnaeus was a botanist who moved to Uppsala in 1741 and became a friend of Celsius. He ordered a 100-degree thermometer from Ekström in 1745, the year after Celsius' death, specifying that zero be the freezing point and 100 the boiling point. Years later, in 1758, Linnaeus asserted in a private letter that he was the inventor of that scale. The other name to mention is Jean Pierre Christin, a scientist in Lyon, France, and a Celsius contemporary. He made

100-degree thermometers, with zero at the freezing point and 100 at the boiling point. The Christin thermometers gained limited acceptance in the area around Lyon and a few locales in southern France, but they never found wide usage. Historians have yet to resolve who should have credit for inverting the 100-degree scale to be the one that now carries the name of Celsius.

That may not have been even an intriguing question until 1948. That was the year the CGPM, the international body that oversees the BIPM, formally attached the name "Celsius" to the 100-degree scale now in common usage worldwide. In doing so, the CGPM abandoned the name "centigrade," which had been commonly used for about two hundred years, because in French the word "grade" is an angular measurement, not a temperature measurement. The 1948 name change did not require any change in the symbol °C, which was already in common usage. Additionally, it was supported by a chemistry textbook which in several editions had been widely-used in Germany during the first half of the 19th century. Rightly or wrongly, that textbook had attributed the 100-degree scale, with zero at freezing and 100 at boiling, to Celsius.

Kelvin Scale

6.4 Lord Kelvin
(William Thomson)

William Thomson (1824 – 1907) was a Scots-Irish physicist and engineer who, as a professor at the University of Glasgow, worked in the field of thermodynamics. He became Lord Kelvin in 1892, making him the first British scientist in the House of Lords. Thomson's work had a broad reach across thermodynamics. It is the attention which he gave to temperature that deserves attention here.

Since early in the 18th century, about 150 years before Thomson began to pay attention, awareness had been building that any liquid in a thermometer, be it water, alcohol, mercury or whatever, does not expand and contract in equal steps up and down the temperature scale. For example, in the common household thermometer, if the temperature on the porch in the winter warms from 20°F to 30°F, the line of alcohol in the instrument might expand up the stem by an assumed 7.00 percent. In the summer, however, if the temperature on the porch warms 10°F again but from 70°F to 80°F, the same alcohol in the same thermometer might expand upward by only an assumed 6.75 percent. Alcohol does not expand linearly, which is to say not uniformly across all ranges of temperature. The degrees marked on the thermometer's scale are uniformly spaced, but the alcohol in the thermometer does not exactly track that spacing. This behavior is not troublesome in a household thermometer where minor inaccuracies are harmless, but it can cause fits in a science laboratory where precision is essential.

This caught Thomson's inquisitive mind. He is credited with being the first to propose the thermodynamic theory that assumes the absence of heat, which is zero temperature, at a common point for all substances. All commotion of particles stops. Temperature is zilch, as is heat. It is a theoretical concept, but it provided a basis for constructing a temperature scale which now provides the precision needed not only in the laboratory but in much of industry.

Thomson, being a scientist with a full appreciation for the metric system, adopted the Celsius scale of 100 degrees, fixing the spread between any two neighboring degrees at the same spread Celsius had given them. Then, as a theoretical exercise, he extended the

Celsius scale backwards down to the assumed point that all heat would be gone. That point was −273.15°C, making it also be zero K (the degree mark ° is not used here, adhering to the BIPM's standard for denoting degrees kelvin).

Thomson's theoretical extension of the Celsius scale to absolute zero left scientists with the challenge of making practical use of it in their laboratories. There was, and there still is, no known material that will behave in a linear manner across all ranges of temperature. The task was to develop a set of practices for the laboratory that in different temperature ranges would give the closest possible approximation to the theoretical scale (a linear line, which is to say a straight line, extending from zero K up to infinity).

More than 100 years after Thomson generated this challenge, the BIPM and national laboratories, including the NIST in the United States, are still puzzling it out. Since 1927, the BIPM has issued a series of publications called International Temperature Scales which have provided recipes for measuring temperatures within consecutive ranges of temperature. By following these recipes, scientists are to be able to make measurements which best approximate the theoretically precise measurements envisioned by Thomson.

The most current set of recipes, adopted in 1990, are the International Temperature Scale of 1990 (ITS-90), now supplemented by the Provisional Low Temperature Scale (PLTS-2000). Needless to say, these recipes are not easy cooking, especially for a non-scientist. Each range of temperature, beginning with the lowest range from 0.009K to 1K, calls for the use of a different measuring substance. None of the recipes involves either mercury or alcohol, the liquids most commonly found in our household thermometers. For the range of temperatures experienced in our everyday lives, the ITS-90 recipe uses platinum resistance thermometers (briefly described below), in conjunction with prescribed mathematical methods of interpolation, to approximate the perfect kelvin reading. We need not look for this up-to-date science along Main Street.

Thermometers, Not Liquid-In-Glass

As already indicated, neither alcohol nor mercury yields the precision often needed in science and industry. More than that, neither of these liquids is suitable for everyday applications where temperature needs to be measured, in some cases continuously monitored, in a safe, stable and convenient manner.

Other designs have been called forth. The basic approach, however, remains the same: first, identify a material, or a combination of materials, that reacts to changes in temperature (that is to say, changes in heat); second, determine with as much accuracy as possible how that reaction manifests itself; and third, harness the material or combination of materials into a device that will measure the reaction. The following paragraphs briefly describe a few of the non-liquid types of thermometers which science has made available, mostly since the beginning of the 20th century.

Thermocouple thermometers are widely used in industry, but they also have important functions in our homes. They might track the room temperature on our thermostat, sending on-off commands to the switches on the furnace and the air conditioner. If the home has a furnace or water heater with a continuously burning pilot light, it might also have a thermocouple which is continuously monitoring the pilot flame while being wired to send a shut off signal on the gas supply if the pilot goes out.

To understand how thermocouples measure temperature, we first need to recall what our science classes taught us about electricity.

Electric current (measured in amperes) carries the oomph. Voltage gives the current a push through the wires. Wires impose a resistance to that push. If the wires are good conductors, such as copper, that resistance is small (the voltage, or push, is only a bit different from one end to the other). If not, that resistance is large (the voltage difference is much bigger), causing the wire to become hot (the wire's resistance converts electrical energy to thermal energy).

Thermocouples take advantage of the observation, first made early in the 19th century, that wires made of different metals react in different ways to the same change in temperature. A thermocouple consists of two wires of different metals which are connected at both ends. At one end the wires are jointly exposed to the target temperature and at the other end they are both held at a known temperature. The thermocouple, which actually measures the voltage difference between the two ends, can then display the target temperature.

Thermocouples use two wires of different metals, but platinum resistance thermometers use a single strand of fine wire, most typically platinum. Like their thermocouple cousins, platinum resistance thermometers do their work by having one end exposed to the target temperature and the other held at a known temperature. When an external power source sends an electrical current through the single strand of wire, the thermometer can display a measurement of target temperature. Platinum resistance thermometers offer higher accuracy and, primarily for that reason, are slowly replacing thermocouples in industrial applications.

Infrared thermometers are devices to measure temperature when the thermometer does not, often cannot, either touch the target (heat conduction) or be touched by a gas or fluid flowing from the target (heat convection). Infrared radiation (heat radiation) provides a third way for a thermometer to "feel" the target's temperature. Expressed more scientifically, an infrared thermometer detects electromagnetic radiation.

This type of thermometer has multiple uses. In medicine, a patient's temperature might be checked by inserting a probe into the ear. The probe has no contact with the inner ear but, instead, functions as an infrared thermometer by picking up heat radiating off the body. Similarly, when a patient or traveler is suspected of having a highly contagious disease such as Ebola, an infrared thermometer might be used to check temperature without having direct contact. Firefighters use infrared thermometers to locate hot spots. Repair personnel use them to detect heat loss zones. Scientists use them to monitor volcanoes and aim telescopes. And, of course, industry uses them in multiple ways.

Recall that heat within a target is evidence of the ongoing commotion among its particles. That commotion means the particles are jumping, spinning and vibrating. When the jumping (scientists call it charge acceleration) and vibrating (dipole oscillation) occur at the target's surface, the particles release heat in the form of thermal energy and some of that energy escapes from the target in radiating waves. The distance between two waves of heat moving away from the target is called wavelength. A target emits countless waves spaced apart in different wavelengths, some more dominant than others.

The human eye can detect only a narrow band of wavelengths. Within that narrow band, our eyes associate different colors with different wavelengths. Viewing a rainbow, the color violet comes from the shortest visible wavelengths of sunlight bouncing off water vapor in the clouds, while the color red comes from the longest. Infrared wavelengths are those that start above visible red at wavelengths of about 1/100ths the width of a human hair. They

> The entire range of potential wavelengths is called the thermal spectrum. Wavelengths that our eyes can detect are within the visible range of the spectrum. Shorter wavelengths (waves more closely spaced) are in the ultraviolent range. Those spaced farther apart are in the infrared range.

continue up the thermal spectrum to wavelengths of about 13 widths of human hair.

A campfire emits a wide range of wavelengths. The human eye can see the yellows, the reds and other colors in the fire; these are the colors our brains associate with wavelengths that are in the visible range of the thermal spectrum. The human eye cannot see and makes no color association with the infrared wavelengths, or for that matter with any of the wavelengths outside the visible range. On a chilly evening the human body when turned toward the campfire can feel its warmth, making an appreciated but unrecognized association with the infrared wavelengths coming from the bed of flames. All wavelengths carry heat.

Like the human body, an infrared thermometer can "feel" the heat in wavelengths that fall within the thermometer's operating range. When pointed at the target with no obstructions to interfere, the infrared thermometer captures a sample of the wavelengths coming from the target and focuses those wavelengths (thermal energy, or heat) on a heat measuring device inside the thermometer, perhaps a thermocouple or platinum resistance thermometer.

Science Marches On

In response to the growing awareness of the toxic hazards with mercury thermometers, scientists in the NIST's thermodynamic metrology group are conducting research to find alternatives. Other

NIST scientists are experimenting with the speed of sound in argon gas to determine whether it offers a more accurate method to measure temperatures in the range of 273K (almost 0°C) to 730K (457°C). In Britain, NPL scientists are looking for improved methods of thermal imaging measurement. In China, at the Chinese Academy of Sciences, an academic body, a team has announced its development of light-emitting particles, so-called nano-thermometers, which are said to be capable of measuring temperature within individual body cells. That development is thought to have promise for the treatment of cancer.

The mind-numbing list of scientific pursuits at the BIPM, at national laboratories around the globe, and in other laboratories, both in academia and industry, could be made much longer. Without doing so here, the point is that scientists at these laboratories are doing sophisticated work, continuing to push the science of thermometry ahead on a global scale.

The development of thermometers, the devices we use to measure the commotion of particles (that commotion signals *heat*, a phenomenon we cannot see for itself), did not take hold until the 17th and 18th century, relatively late in the scheme of things. The Age of Enlightenment brought a surge of interest in how things work along with an active pursuit of ways to make life better. Across Europe several curious and inventive persons dedicated themselves to improving and making thermometers, motivated in different degrees by curiosity, market demand, and profit.

Among the many possibilities, three types of thermometers survived the competition. Fahrenheit thermometers became a widely-accepted device in the 18th century. Soon thereafter, Celsius thermometers, using a metric scale, became a new standard and over the next two centuries displaced Fahrenheit in many places (a big exception is the United States, where Fahrenheit still prevails). During the 19th century Lord Kelvin, a Scottish scientist who studied heat and methods of its measurement, continued the development of the Celsius thermometer. In 1954 the BIPM made temperature one of its basic units and in 1967 chose the word "kelvin" to designate one unit of heat. Now, in the 21st century, research is ongoing to meet the demands, primarily in laboratories, for ever more accurate measurements of heat, up and down an increasing range of temperature.

Chapter Seven
CLIMATE

During Flight 1230's descent into Beijing Capital International Airport, at the end of a long but routine flight from Washington-Dulles, passengers seated near the windows saw a gray blanket of smog covering the Beijing metropolitan area. Many were not surprised, having seen and heard international press reports about China's efforts to come to grips with air pollution.

The passenger in seat 34A took a special interest. She was an employee of NASA who had been involved for several years in climate studies. The focus of her work had been weather patterns across the United States, and that work had brought her to a viewpoint that climate in the United States can only be understood as a sub-topic of climate around the globe. She had firmly fixed in her mind the often-quoted question posed at a 1979 meeting of scientists: "Does the flap of a butterfly's wings in Brazil set off a tornado in Texas?"

This American scientist had come to Beijing, on the invitation of the Chinese government, to consult with environmental officials and their advisors about steps taken in the

United States to reduce air pollution. She would have both successes and failures to share with them, along with many not-yet-answered questions.

One Planet

The pollutants in the smog over Beijing will not leave the planet, they will just disperse to other locations around the globe. The same is true, of course, for pollutants that rise from areas in the United States and other countries. In varying types and levels around the globe, pollutants are in the air we all breathe. And there are other ways, less obvious than observable pollutants, that the climate affects every one of us. Climate is a very big part of everyday life, globally.

> Climate expresses itself in today's weather. A change in the weather, however, says next-to-nothing about the climate. A change in the climate says a great deal about the weather, not just for the here and today but for the entire globe and for the years, decades and centuries to come.

Recall the German astronaut on the international space satellite who has told us that our planet has a thin, fragile atmosphere and that we are one humanity, all sharing the same fate. Climate scientists, or nearly all of them (the oft-cited count is 97 percent), are telling us the

> "Climate change" and "global warming" are two different concepts, but they are causally related. Global warming is the long-term trend of rising average temperature. Climate change refers to consequences of global warming, such as precipitation patterns, heat waves, and weather extremes. Much of the literature uses the two terms interchangeably.

same thing. More than that, they are telling us, with increasing certainty, backed by increasing data, that our thin, fragile climate is changing much more rapidly than it ever has in the planet's long past and that some part of that rapid change is caused by human activities. Other scientists (the three percent) disagree. But nearly all climate scientists should be able to agree that neither viewpoint is imbued with absolute certainty. Solid science chases after certainty, but never catches it.

When Arabic numerals were brought to Spain in the 10th century, it took several more centuries for them to gain widespread usage in other parts of Europe. When the Pope in the 16th century decreed the use of the Gregorian calendar, it took the Protestant population several more centuries to accept the Pope's correction. When French revolutionists late in the 18th century switched that nation to the metric system of measurement, it took the French population about 50 years to fall into line and the British more than 150 years. Many British measurements and even more United States' measurements are still pinned to non-metric units. China did not begin its push into metric until the last half of the 20th century. Earlier chapters in this book have mentioned other changes in everyday life that have come about only after much delay. Change can be tough stuff. It requires people to unstick themselves from long-held habits and attitudes. Change often comes only after its benefits have become undeniably obvious, even to those who begrudge the disruption it brings to their everyday lives.

Climate change is different. Delay in acting to control global warming could very well not be a feasible option. As we go about our everyday lives, we are unwittingly adding to global warming, while scientists are telling us we are upping the risks that global warming imposes on our planet. They tell us that nature has its own points of no-return and we may well not foresee those tipping points before it becomes too late. But still, many of us take note of climate change only when extreme weather happens to come our way. In our daily activities we generally give thought to how we can best use our time, we generally give thought to how our own behavior might affect others, and we generally give thought to expenses we should expect to incur. In contrast, few of us give much, if any, thought to how our daily lives might affect the climate in which we and others are living. For many of us, the sources from which we draw energy and the impacts those sources impose on the Earth's climate are too remote to take up space in our busy lives. Even if we have a concern about climate change over the long haul, the time scale for impacts that might affect us individually is unknown. Moreover, as single beings in a vast and rapidly growing population around the globe, how can the teensy, tiny bits of energy we use in our own daily lives possibly have an impact on the entire globe? It is too easy to tell ourselves that if we must concern ourselves with climate change we will do so tomorrow, not today. For now, we are too much content to let policymakers wrestle with the problem. That must change. Each of us needs to give our climate the same attention we give to the use of our time, the effects of our behavior on other people and the expenses we incur.

Without recognizing it, we are governed by the climate in many ways. In our everyday lives we are breathing air, feeling temperature, enjoying sunlight (or not), splashing rain puddles (or not), frolicking in snow (or not), or experiencing in some other way the weather events which the climate sends our way. Climate affects crops. It affects ocean levels and currents, which in turn affect coastal waters. It affects not just humans but all living species, many of which are too small or too remote to be given much notice. All of these phenomena and others express themselves in our everyday lives. When they change, our lives change.

Backstory

Wine producers see a cause and effect linkage between the climate where their grapes are grown and the characteristics of what goes into their bottles. Some annual vintages excel, others are just so-so. That same poetic association of climate with outcome was at the root of explanations that were given, perhaps well into the 20th century, for making a linkage between climate and prosperity. Global warming was not a concern. Moderation of weather extremes held the attention.

In Europe during the 17th and 18th centuries, many whom we now classify as Enlightenment philosophers saw a connection between climate and human advancement. During and after the Middle Ages, forests in Europe had been cleared, additional lands had been cultivated and prosperity had spread across a growing population. These gradual changes had come about because the continent was in a favored latitude with a climate which, moderated by human efforts, had in turn fostered strong characteristics of mind and body. Weather extremes could be attributed in some part to God's providence.

Across the Atlantic, the Americas were being settled by colonists, many of whom found themselves in a surprisingly harsh climate. Their challenge was to tame the new land by draining marshes, clearing forests, and cultivating lands. Their vision was to transform the new-found territory, along with its climate, into a new land promising widespread prosperity and good health, as had been the perceived experience in Europe.

As might be expected, on both sides of the Atlantic contrary voices were heard, denying any linkage between climate and prosperity.

For these naysayers, the evidence was just not sufficient. But proponents of the linkage were at the forefront of establishing weather data systems which, in their expectations, would validate their viewpoint. The road was not smooth, but during the 18th and 19th centuries both private and public weather systems came onto the scene in both Europe and the Americas. Although not their intended purpose, over time these systems were to become essential sources of data when physical science, rather than philosophy, would take the lead in trying to understand our climate.

Fairness requires us to acknowledge that thoughtful persons before and into the industrial age were doing the best they could, with available knowledge, to fit our planet's climate into a rational framework. And it is to be noted, before proceeding to the story of science and climate, that their philosophical approach put a finger on one enduring concept. As science has now determined, human activity does indeed affect climate.

Scientific attention to our planet's atmosphere did not take a firm hold until early in the 19th century. And then it took about another 100 years or so to generate an awareness that the Earth's climate is undergoing permanent change, due in some part to human causes. That gives the climate change story a very short timeline, when compared with the much longer timelines we have seen in this book about other fundamentals of our everyday lives. The BIPM does not have a front-line role in climate science, but it has a critical role behind the scenes in that the units of measurement it establishes allow climate scientists around the globe, in their many different roles, to communicate effectively with one another.

The story began in France in the early decades of the 19th century, when Jean-Batiste Joseph Fourier (1768 – 1830) calculated that the Earth, at its distance from the Sun, should be much colder than it is.

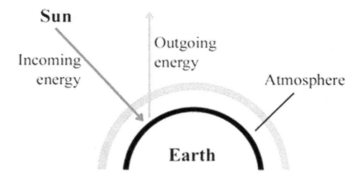

7.1 Greenhouse Effect   Gases accumulate in the Earth's atmosphere, trapping outgoing energy and causing the climate to warm.

("Fourier" is a familiar name among scientists and mathematicians. Much of Fourier's work dealt with the flow of heat, and his name is attached to mathematical tools for analyzing that concept.) He reasoned that infrared radiation from the Sun, having been reflected off the Earth's surface, could not easily escape the atmosphere and that the warmer air it produced was acting as an insulating blanket. Fourier did not live long enough to learn that he would be deemed by many to have been the first to recognize what we now call the "greenhouse effect."

What was in that insulating blanket? About 40 years later, John Tyndall (1820 – 1893), both an English historian and a pioneer in Alpine climbing, determined that northern Europe had once been covered by ice sheets. That led him to the discovery that both water vapor and carbon dioxide (CO_2) trap heat. Oxygen and nitrogen, the main gases in the atmosphere, do not. The discovery by Tyndall was advanced, around the turn of the century, when Svante Arrhenius (1859 – 1927), a Swedish scientist, reasoned that the level of water vapor in the atmosphere fluctuates but CO_2 is a much longer lasting component. He also concluded that warmer air can hold more water vapor and when water vapor is present it causes even more warming. (Scientists call that positive feedback.) Arrhenius did his work in the final decade of the 19th century, when there was little awareness that the consumption of fossil fuels (coal, oil and gas) would accelerate as rapidly as it did in the decades that were to follow. The world economy was humming. No great risk was foreseen.

Jean Batiste Fourier   John Tyndall   Svante Arrenhaus

7.2 The so-called grandfathers of climate science

The first half of the 20th century has been called a "sleepy backwater" for climate science. It brought efforts by a few in the science community to revisit and refine the 19th century investigations. They were hampered by imperfections in laboratory devices, overestimations of the capacities of oceans and terrestrial plants to absorb CO_2, and an oversimplified understanding of the atmosphere itself. A common view was that the climate is a self-regulating mechanism that will cope with any influx of carbon from fossil fuels. Throughout the early 20th century, the primary interest among scientists who were paying attention to the Earth's climate was to probe the mystery of the early ice ages. The here and now of the Earth's climate had only secondary importance. Of even less interest was its future.

7.3 Guy Stewart Callendar

It was not until post-World War II that the insulating blanket began to attract wider attention. The work of Guy Stewart Callendar (1897 – 1964), an English steam engineer by day who followed meteorology as a hobby, is credited with keeping interest in global warming alive. His investigations and writings during the 1930s and continuing into the 1950s advocated that the Earth was warming and, in the long term, that warming could influence the climate. He put his finger on industrial emissions, but the rate of global warming was still not high enough to sound an alarm.

With the advent of the Cold War, governmental funding for the development of weapons systems flowed into scientific investigations, a good part of which had to do with developing a better understanding of the climate and how it affects navigation and targeting. Scientific tools were becoming more accurate. At the same time, computers were coming on the scene and their powers of computation presented an enticement for the development of digital models which could digest as many details about the climate as might be fed into them.

From about 1950 forward, climate studies expanded in scope to become a multi-disciplinary field. Atmospheric scientists were joined by biologists, geochemists, oceanographers, computer specialists and others. It took years, however, for the many experts in different branches of science to bring their efforts together under the umbrella of climate science. Among other occasions, a conference titled "Causes of Climate Change," held in Boulder, Colorado, in August 1965, has been tagged to have been an eye-opener for scientists of different disciplines. They had gathered to develop a shared understanding about the complexities of the climate and how their seemingly diverse interests were in fact interwoven.

Ice cores taken from Greenland and the Antarctic allowed scientists to find a close correlation between CO_2 levels in the atmosphere and temperatures at the surface, over long stretches in the Earth's past. The most notable ice cores have been extracted in the Antarctic at the Vostok and Dome C stations. The Vostok core, extracted in 1999, has yielded measurements of gases trapped about 400,000 years ago, while measurements from the Dome C core, extracted a few years later, reach back about 800,000 years. Over these long time periods, embracing multiple ice ages, the levels of CO_2 in the atmosphere ranged between 170 and 300 parts per million (ppm).

Importantly, the periods of highest CO_2 levels correlate, but do not coincide, with periods of warmer global temperatures. Ice cores have shown that the buildup of CO_2 lagged by several centuries the start of atmospheric warming. (That confirms earlier studies indicating that ice ages are triggered by small shifts in the Earth's orbit around the Sun, the so-called "Milankovitch cycles".) Once a shift has initiated a warming, levels of CO_2 in the atmosphere begin to rise gradually and then a positive feedback promotes additional warming, causing still more CO_2 to stay in the atmosphere, which causes still more warming, and so on.

Flora and fauna found in fossils going back hundreds of millions of years indicate that warmth-loving species had moved north in warmer periods, while evidence has accumulated that CO_2 levels had been much higher during the same periods. Distinctive rock sequences have yielded clues that the planet in long periods past had been much warmer than it ought to have been, given that the Sun long ago had begun emitting less energy (scientists call this the "Faint Young Sun Paradox").

In 2009, one expert on climate science complained that CO_2 was "sucking all of the oxygen out of the room." Behind this complaint was his assessment that public debate had been too much ignoring other gases that are known to be additional contributors to the greenhouse effect. Water vapor has long been recognized. Chief among other gases is methane (the main ingredient of natural gas), which flows into the atmosphere primarily from wetlands and rice paddies. The melting of Arctic tundra carries the potential for future eruptions of methane trapped in the permafrost of its terrain. Leaking oil and gas pipelines and burping cows also play their parts. Other contributing gases include nitrous oxide (primarily released by fertilizers) and chlorofluorocarbons (CFCs, releases of which have been discouraged to protect ozone in the atmosphere). These contributing gases, on a molecule-by-molecule basis, even if not on an overall basis, can have a greenhouse effect many times greater than CO_2.

Collectively, all the gases that contribute to the greenhouse effect are lumped together as "greenhouse gases." Scientists estimate that CO_2 emissions account for a bit more than half of the effect that humans have on climate change. The remainder is attributed to other greenhouse gas emissions.

Ocean's Role

The ocean covers about 70 percent of the Earth's surface and holds about 97 percent of its water. The majority of the sunlight that reaches the Earth lands on the ocean, which means that the water soaks up a great deal of the energy coming from the Sun. A pause in ocean surface warming can occur for various reasons, including La Niña events in the Pacific (explained below) and volcanic eruptions, but a pause at the surface does not mean the ocean, when considered throughout its depth, and hence the Earth as a whole, is not continuing to warm. The top few meters of the ocean store about as much heat as the Earth's entire atmosphere, and much of that heat is stirred down into deeper depths. According to latest estimates, more than 90 percent of human-caused warming goes into the ocean. The heat absorbed by the ocean sustains its food chain, starting with algae and plankton and progressing up the chain to fish, whales and seals.

> Navigators and map makers put identifying labels (Atlantic, Pacific, and so forth) on different regions of the Earth's ocean, but they all connect to be one big body of water. Climate scientists treat the ocean as one big body.

The ocean is continuously interacting with both the Earth's atmosphere and its land masses (including forests and other plant habitats that both absorb and emit CO_2). Collectively, the three together are like a huge jigsaw puzzle for climate scientists who are trying to piece it together. They have much yet to do. With many questions still unanswered and, almost certainly, many not yet even asked, it has nevertheless become clear to those who are probing the climate's complexities that the ocean does indeed have a major role. Four pieces are holding most of their attention.

> El Niño (in Spanish, the little boy, or Christ child) and La Niña (the little girl) refer to warm and cold phases of ocean water in the east-central area of the Pacific. El Niño, the warm water phase, typically brings warmer temperatures to western and northern United States and western Canada, along with more rainfall in the Gulf Coast. La Niña, the cold water phase, typically brings opposite effects. (Find more information at http://earthobservatory.nasa.gov/Features/ElNino/and/LaNina/.)

First, ocean water at the surface is warmer in the tropical zones, where sunlight more consistently strikes the Earth from directly overhead and, therefore, is less likely to be reflected away. That means evaporation is more prevalent in tropical regions, and that in turn means more rainfall reaches land there.

Second, differences in temperature at the Earth's surface, both land and ocean, create wind currents while differences in temperature and salinity in the ocean create water currents. Adding to that, wind currents over the ocean generate waves which further promote water currents. These forces show themselves in what scientists have come to call the Great Ocean Conveyor Belt. That globe-wrapping belt is the dominant pattern of warm water movements from the equator to the poles, followed at deeper depths by colder and saltier water movements back toward the equator. The ocean is a super-super-sized mixing bowl.

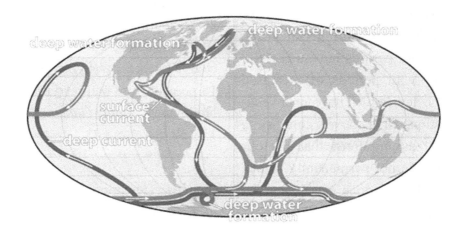

7.4 Great Ocean Conveyor Belt Major ocean currents carry warmer surface water from the equator to the poles and deeper colder water back towards the equator.

Third, the ocean absorbs about one-fourth of the CO_2 emitted into the atmosphere from the burning of fossil fuels. That absorbed CO_2 makes the ocean acidic. At some point, the acidity interferes with the capacity of ocean creatures like mussels and coral to make strong protective shells.

Fourth, salt in the ocean affects its currents. Salty ocean water is heavier and tends to sink. Glaciers and icebergs do not contain salt. When they melt and become ocean water, the saltiness of the surrounding water is changed. Scientist are asking what effects that melting might have on other consequences of global warming.

Today's Climate

Overall, evidence has been accumulating from multiple directions that the Earth's climate has been fluctuating between periods of warmth and cold for millions and millions of years. Warmer periods have a correlation with periods of increased CO_2 levels. Moreover, investigations have shown, contrary to past assumptions, that the climate is not permanently stable.

> The primary source of CO_2 emissions is the burning of fossil fuels (coal, oil, and natural gas). Globally, for 2010, about 42 percent of CO_2 emissions came from coal, 34 percent from oil, and 18 percent from natural gas. Nordhaus, *Climate Casino*, 158-59. See Note 23.

Small variations can trigger large changes. The important difference between the distant past and the here and now is the rate of change. Before the industrial revolution (roughly, before 1750), the level of CO_2 in the atmosphere was about 280 parts per million. That concentration climbed throughout the rest of the 18th century and then throughout the 19th and the 20th centuries. During the 20th

Climate

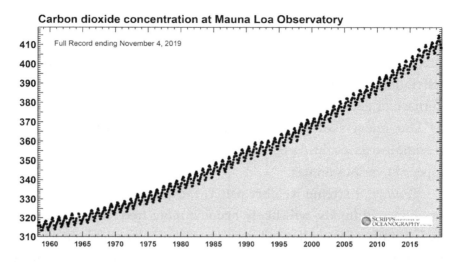

7.5 Carbon dioxide in the Earth's atmosphere

century, both the world's population and energy usage per person quadrupled, bringing a sixteen-fold increase in the rate of CO_2 emissions. Today's levels exceed 400 parts per million, a level that last occurred three million years ago. On the current trend, CO_2 would double its preindustrial level within 50 to 60 years, putting it at 560 parts per million well before the end of the 21st century.

This global data can be brought into our own lives. The amounts of CO_2 emitted by burning fossil fuels, per million British thermal units (BTUs) of energy produced, are roughly 215 pounds for coal, 160 for oil, and 117 for gas. One lone BTU, a common measurement of heat (thermal energy), raises the temperature of one pound of water by one degree Fahrenheit. That is not much heat. But an average household in the United States consumes a whole bunch of these little bits. Survey data compiled by the United States Energy Information Administration indicates that in 2009 the average household used about 90 million BTUs. Combining that data, the average household in 2009 put about 19,350 pounds (approaching

10 tons) of CO_2 into the atmosphere if its energy came entirely from coal (burned at a distant generating station), 14,400 pounds (seven tons) if from oil (burned in the home furnace and/or at a distant generating station), or 10,530 pounds (over five tons) if from gas (furnace or generating station).

The impacts of increasing concentrations of CO_2 and other greenhouse gases, bringing more global warming, are forecasted to appear in various arenas:

Weather: Extreme weather patterns, especially those that bring droughts and floods, will likely become more frequent. Those who study weather are challenged to peg any particular event upon global warming, but they can make statistical observations that relate weather *patterns* to global warming. ("Attribution science," which attempts to connect weather events to climate change, is a relatively new front for scientists studying climate change.) The devastation in New Jersey, New York and Connecticut in 2012, when Superstorm Sandy hit those coastal areas, has given climate scientists good reason to put even more attention upon the risks that global warming is bringing to coastal areas. On a wider scale, heat waves, droughts, heavy rainfalls and snowstorms, increases in the frequency and severity of storms, tornadoes, hurricanes and other regional and local weather occurrences are all likewise drawing attention.

Food supplies: Contrary to popular rhetoric, global food production is projected to *increase* as the average temperature increases over a range of 1° to 3°C. Within that temperature band, however, changes in precipitation patterns will cause major local disruptions. The regions thought to be most vulnerable are Africa and Western Asia. Once that range of temperature is exceeded, hundreds of millions of people will be at risk of hunger due to failing crops.

Water supplies: Mountain glaciers and winter snowpacks are shrinking at increasing rates, jeopardizing water supply systems.

Sea levels: The two major ice sheets on the planet are those of Greenland and West Antarctic. Ice sheet melting is fraught with the risk of unpredictable tipping points. A study released by the National Academy of Sciences concludes that the rise of the of the planet's sea level is accelerating and that half of this acceleration is driven by melting ice in the Antarctic and Greenland. Meanwhile, storm surges will cause local emergencies. Coastal cities are already seeing this impact.

Ocean currents: The Gulf Stream conveys warm surface water into the North Atlantic, allowing for temperatures in Britain and other northern European countries to be warmer than their latitudes would indicate. Large shifts of direction, however, occurred during earlier times, especially during the ice ages. Warmer surface temperatures and related changes in precipitation patterns could alter the speed or even the direction of the Gulf Stream, bringing unforeseeable consequences to those areas now relying upon it for their climatic conditions. Other ocean currents could be subject to similar risks.

Species survival: Plants and animals that make their habitats on land will be forced to adjust to changing conditions. Rising CO_2 concentrations in the atmosphere lead to rising concentrations in the ocean, making the ocean more acidic and lowering their concentrations of calcium carbonate. This puts at risk those marine organisms that form shells, such as corals, mollusks, crustaceans, and some plankton. Some species may be able to shift their habitats, but others may become extinct.

Human conflicts: Weather extremes, food and water shortages, coastal floods, and other direct consequences of global warming stimulate and amplify human migrations. The conflicts often triggered by mixing populations with different cultures then bring violence between those groups, which in turn induces other populations to become involved. Indeed, the refugees who have been displaced from their historical homelands in Syria, Iraq and other Middle Eastern nations in recent years are widely thought to have been victims of shortages caused by regional drought. Their efforts to find refuge have in many cases triggered violence and warfare, not only in their native lands but in neighboring areas as well.

Some of these effects of climate change can generate feedbacks which either accelerate the primary effect or launch secondary ones. This raises the risk of trigger points which, once reached, make the process irreversible. Little is known about what those trigger points are, and that uncertainty presents one of the greatest risks. A 2016 event of "river piracy" in the Canadian Yukon dramatizes the uncertainty. Glacial meltwater changed its historical direction of flow, in what climate scientists call a "geological instant," by redirecting itself from a northbound river path towards the Bering Sea to a southbound path towards the Pacific Ocean. Similar events can happen anywhere around the globe, and they can have life-changing impacts on both humans and wildlife.

As scientists have been learning more and more about the insulating blanket that Fourier detected 200 years ago, they have come to understand that the climate is a boiling stew of many ingredients, made even more complex by multiple feedback loops. How do phenomena on land affect the ocean? How do phenomena in the ocean affect land? As each question is asked new questions appear. And as the complexity has grown, so has the difficulty of puzzling out all the causes of global warming and the multiple effects of that warming on our planet. Even if they cannot agree fully on what causes what, nearly all scientists (the 97 per cent) do agree, first, that the climate is indeed warming at an unprecedented rate and, second, that there is convincing evidence that human activities are to some degree a cause, most likely a significant cause. Like their colleagues in other arenas, climate scientists are pushing forward on a global scale. There is much still to be explored. Satellite images are helping to fit pieces together, while at the same time those images are revealing more to be considered. Like other scientists, those concentrating on the climate are finding their own challenges.

Public Awareness

Public interest in the environment and concern that humans were neglecting what they should instead be protecting was aroused with the publication in 1962 of the bestseller *Silent Spring,* by Rachel Carson (1907-1964), an American naturalist turned author. That widely read and highly praised book put a wrapper of moral responsibility around environmental protection. Growing public concern about urban smog, especially in the Los Angeles area, underscored what Carson was saying.

Carson's eye-opening book was followed in 1992 with the publication of *Earth in the Balance,* by Al Gore (1948 −). Gore was then a member of the Senate and later that year would be elected to become Vice-President of the United States. Building public awareness about climate change had become one of his primary missions. In 2006, Gore's book became the basis for a documentary film titled *An Inconvenient Truth,* and that production was quickly followed in 2007 by a republication of Gore's 1992 book, this time carrying the same title as the film. Public attention zoomed.

In June 2015, six months before national leaders were to meet in Paris to make yet another pass at an international agreement about climate change, Pope Francis released his encyclical, arguing for a partnership between science and religion to confront the issue. His stance was based on social justice, and he put an emphasis upon the burdens that populations in developing countries would bear if indifference about climate degradation were to continue. The Pope's voice was heard, by both activists and skeptics.

School curricula, newspapers and magazines, television, the internet, library and bookstore shelves and other places where we all try to keep ourselves abreast of current topics are heavily populated with information and viewpoints about climate change. The debate is now very much in the public square. Public media, with increasing frequency, is bringing attention to environmental events and observations around the globe which most climate scientists and many others see to be evidence of climate change. Increasingly, humankind is coming to a recognition that we have been living too long on a binge of inexpensive fossil fuels and the time has come to sober up. But there are others among us who are much less concerned, while they continue to see the climate we all share still to be in line with historical patterns of natural variability.

State and National Policies

Among the many things that can arouse public interest, changes in the automobiles we drive is right up there at the top. California lawmakers, confronted with Los Angeles smog in the 1950s and 1960s, plowed the first furrows into what has since become a burgeoning field of environmental regulation. Vehicle emission controls first took effect in California for all new 1961-model cars, and within about three years similar controls had spread across the United States.

The United States Environmental Protection Agency (EPA) was established in 1970. Regulatory bodies were also established in Canada, Western Europe, Japan and Australia. In the years that followed, as more was learned about vehicle emissions, additional controls came into force. Cars, a consuming interest for many, took climate change from the scientists' laboratories and drove it into the public conscience. Aside from vehicle emissions, other hazards touching the climate and the environment generally were coming into much sharper focus during the 1960s and the decades that followed. Water, industrial stacks, waste disposal, pesticides, fertilizers, insecticides, radioactive materials, and other potential trouble spots have come increasingly into the public spotlight.

In the United States, the EPA has become the chief national rule maker and head cop for all things that affect the environment, not the least of which is the climate. In Britain, the constituent governments of England, Scotland, Wales and Northern Ireland each has its own environmental regulatory agency. In the EU, the European Environment Agency functions as an information source for the EU's member nations, each of which addresses the climate and other environmental concerns through its own governmental procedures.

Flight 1230's destination was China. That nation has 16 of the world's 20 most polluted cities, a consequence generally attributed to its rapid economic growth over the past several decades. Environmental policy in China is set by the national legislature, the National People's Congress, and managed by the Ministry of Environmental Protection. The central government issues regulations but monitoring and enforcement are largely left to local governments. Within a system that traditionally has rewarded economic growth, without much regard for environmental impacts, aligning local enforcement efforts with national environmental policies has been a struggle. In recent years, however, citizen activism about environmental issues has increased.

In 2014, the National People's Congress adopted a more rigorous environmental policy for China, prompted in no small way by the pollution levels plaguing Beijing and other large urban areas. Taking effect in 2015, that updated law specifically addresses air pollution and calls for additional government oversight. Significantly, local governments have been made subject to discipline for failing to enforce environmental laws, and regions are to be judged on how well they balance economic performance with environmental protection. Time will tell.

International Efforts

International cooperation comes grudgingly and gradually. Every nation has its own problems and its own interests. Compounding that, every nation's leaders have their own perspectives. The history of international conferences, the list of compromising pronouncements, and the list of international bodies that must come together to reach any understanding that binds the whole all stand as evidence that national sovereignty, in combination with national politics, is very much in the way of international problem solving about the climate.

But the thing about climate change is that it makes no accommodation for the sovereignty or politics of nations. Borders have no sanctity. Humankind is living in a single climate worldwide, and climate change is happening worldwide. Changes of a no-going-back nature are taking place while conferences are being scheduled, compromises are being negotiated, timelines are being extended, and so forth and so on. Even a halfway complete history of international efforts to address climate change would take a reader into an alphabet soup of official names and acronyms denoting the many organizations, the many meetings, the many pronouncements, and the many other events that tell a complex and largely discouraging story. Only a few need to be mentioned to capture the essence. That history is still playing out. The window of time for some breeze of resolution to come along is probably still open, but that is far from certain.

The first World Climate Conference, organized by the World Meteorological Organization (WMO), was held in Geneva in 1979. It was primarily a meeting for scientists, not diplomats or politicians. That 1979 conference led to the establishment in 1988, by the WMO and the United Nations Environment Program (UNEP), of the Intergovernmental Panel on Climate Change (IPCC). In 1990 the IPCC called for a global treaty on climate change, and that treaty was adopted in 1992 by over one hundred nations at a conference in Rio de Janeiro. The treaty is officially called the United Nations Framework Conference on Climate Change (UNFCCC). The stated objective of the UNFCCC is to "stabilize greenhouse gas concentrations in the atmosphere at a level which would prevent dangerous anthropogenic interference with the climate system." In the alphabet soup, nations that are parties to the UNFCCC are "Parties to the Convention" and they meet from time to time at "Conferences of the Parties," or "COPs." Do not ask, just eat the soup.

The Kyoto Protocol was adopted in 1997, at COP3, to augment the terms of the UNFCCC by setting specific percentages for the reduction of greenhouse gas emissions. Each nation that committed itself to the Protocol was assigned a specific goal. The goal assigned to the United States was to reduce its emissions by seven percent below its 1990 level, and that reduction was to be achieved by the end of the Protocol's first commitment period 2008 to 2012. The reduction percentages were to increase for all participating nations in the second commitment period 2013 to 2020.

The United States signed the Protocol in 1998, but Senate ratification was never requested or obtained. Canada, Japan, Russia and several other countries were parties through the first commitment period, but then they withdrew without making any reduction commitments for the 2013 to 2020 period. As of mid-2016, the Kyoto Protocol was dangling on a thin string with only 37 nations committed to binding emission reductions, and with only seven of those having ratified their commitments.

The 15th Conference of the Parties, or COP15, was held in Copenhagen in 2009. Hopes ran high that commitments would be made to reduce emissions, and about 115 world leaders attended the closing segment with those hopes in mind. But the outcome fell far short of expectations. The so-called Accord included the long-term goal of limiting the maximum global average temperature increase to no more than two degrees Celsius above pre-industrial levels, with that goal to be reviewed in 2015. But, and this is the big but, there was no agreement on how to do this. The word "Accord" is a stretch.

The Copenhagen disappointment stiffened the determination to try again. Careful preparations were made in anticipation of the next effort scheduled for 2015. In 2014 the European Union committed itself to double its renewable energy to 27 percent by 2023, and in November of that year the two leaders of the United States and China jointly committed their nations to cuts in carbon emissions. Concurrently, the IPCC late in 2014 released the final part of its Fifth Assessment Report, providing an up-to-date scientific view of climate change. Then in December 2014 the capstone of the preparatory efforts for the forthcoming 2015 effort was an international conference in Lima, the purpose and outcome of which included the negotiation of a draft agreement under which all participating nations would contribute to a low carbon future. Significantly, Lima called for all nations to come forth ahead of the Paris gathering with their own intended commitments for reducing emissions of greenhouse gases.

December 2015, á Paris! COP21 was attended not only by the

scientists who had first-hand knowledge of the issues but also by national leaders who came with their own climate control commitments already on the table. President Obama of the United States and Premier Xi Jinping of China assumed key roles, wanting to build upon their 2014 bilateral commitment to curb carbon emissions and to set an example for other world leaders.

The Paris Agreement, signed by 197 nations, itself offers no long-term solutions. Instead, it commits those nations who stay with it to come forth in the years ahead with their own methods for addressing the problem. Its central objective is to hold the increase in global average temperature to 3.6°F (2°C) above pre-industrial levels, while its aspirational objective is 2.7°F (1.5°C). Addressing those objectives, the Climate Science Special Report published by the United States government in 2017 states:

> *Achieving global greenhouse emissions reductions before 2030 consistent with targets announced by governments in the lead up to the Paris climate conference would hold open the possibility of meeting the long-term temperature goal of limiting global warming to 3.6°F (2°C) above pre-industrial levels, whereas there would be virtually no chance if net global emissions followed a pathway well above those implied by country announcements. <u>Actions in the announcements are, by themselves, insufficient to meet a 3.6°F (2°C) goal</u>; the likelihood of achieving that depends strongly on the magnitude of global reductions after 2030. [Emphasis added.]*

Global temperature records start around 1880, long after the start of industrialization. Since 1880 the average global temperature has increased about 0.8°C, and about two-thirds of that increase has occurred since 1975. In the absence of temperature records that reach back to the start of industrialization, we are unlikely ever to know whether the 2°C limit of the Paris Agreement has been met.

Coming as it did on the heels of more than three decades of intermittent conferences which had brought much discussion but no small amount of blame-casting and little commitment, the Paris Agreement can be seen, admittedly through a wide lens of optimism, to be a landmark of international progress. The community of nations that signed it were together saying that yes, all of us on this planet are sharing a single global climate. We should all share the responsibility to take care of it.

After Paris

A major setback has occurred. President Trump on June 1, 2017, less than five months after taking office, announced that the United States would withdraw from the Paris Agreement and immediately cease all implementation of its terms. Formal notice of withdrawal was delivered to the United Nations on November 4, 2019, the first permissible date, and is to become effective one year later. Interestingly, in November 2017, five months after President Trump's 2017 announcement, the United States government released its Climate Science Special Report setting forth information about environmental challenges and concluding that "it is extremely likely that human influence has been the dominant cause of observed warming since the mid-20th century." Additionally, in spite of President Trump's 2017 announcement, the United States in December 2018 sent a delegation to the follow-up conference in Katowice, Poland, at which a set of rules was adopted for implementing the Paris Agreement, and the United States agreed to those rules (some controversial proposals were postponed for consideration again at a subsequent conference).

The Paris Agreement was an international achievement because it acknowledged climate change to be a global problem, but the Agreement, now in effect, says nothing about how to confront that problem. So the most important conversation about climate change has only begun. Within the community of climate scientists, a so-called "dirty little secret" is attributed to the Paris Agreement: the Agreement does not acknowledge that, to meet its goal of limiting the increase of global surface temperature to 2°C (3.6°F), the *flow* of new CO_2 emissions into the atmosphere, sometime during the 21st century – and sooner rather than later – must not only decline but must turn from positive to negative. More simply stated, *accumulated* CO_2 must be taken out of the atmosphere (negative emissions).

This is a "dirty little secret" because at present there is no proven technology for achieving negative emissions at a scale even close to the challenge that lies ahead. Carbon capture coupled with underground storage has received attention, as have establishment and re-establishment of forests, chemical treatment of terrain to increase absorption of CO_2 and other methods. But all of the technologies which have been proposed for achieving negative emissions at the scale the planet will need carry with them potential economic, political and other disruptive impacts. Climate change might have now been widely acknowledged, but much is yet to be understood and many tough decisions have yet to be made.

While the search for a better understanding continues, those tough decisions are being postponed and our climate, more rapidly than we have been assuming, is becoming less livable under accustomed standards. The report from the UN Summit in September 2019, amid its more positive passages apparently intended to promote a "we can do it" attitude, lays bare the status we are in as we approach the fifth anniversary of the Paris Agreement:

"The Climate Action Summit 2019 reinforced the global understanding that 1.5° C is the socially, economically, politically and scientifically safe limit to global warming by the end of this century, and to achieve this, the world needs to work to achieve net zero emissions by 2050.

There was also widespread concern [at the Summit] that the world is presently way off course to meet the global target, as emissions continue to increase, and global temperatures rise. The last four years [2015 – 2018] were the four hottest on record, and winter temperatures in the Arctic have risen by 3°C since 1990. Sea levels are rising, coral reefs are dying, and we are starting to see the life-threatening impact of climate change on health, through air pollution, heatwaves, and risks to food security.

Standing behind these upsetting observations is the work of the scientists associated with the International Panel on Climate Change (IPCC), which in 2018 released its most recent assessment on the impacts of global warming.

In another section, the report of the UN Summit in 2019 acknowledges that the task of preventing climate change is becoming ever more challenging:

"The UN estimates that the world would need to increase its efforts between three- and five fold to contain climate change to the levels dictated by science – a 1.5°C rise at most – and avoid escalating climate damage already taking place around the world."

The report quotes UN Secretary-General António Guterres:

> *"The best science, according to the Intergovernmental Panel on Climate Change, tells us that any temperature rise above 1.5 degrees centigrade will lead to major and irreversible damage to the ecosystems that support us .... Science tells us that on our current path, we face at least 3-degrees Celsius of global heating by the end of the century.*
>
> *The climate emergency is a race we are losing, but it is a race we can win."*

These quotations from the UN Summit demonstrate that various organizations around the globe which are already concerned about our climate are, with scientifically-based reasons, more than justified to be concerned. Organizations of youth are especially concerned, as evidenced by the first ever Youth Climate Summit which was convened at the UN on September 21, 2019, just before the UN Summit.

The big problem, of course, is that the world's general population and its political leaders are not yet fully on board. In an opinion prompted by the fires in Australia but aimed at the worldwide population, the Nobel prize winning economist Paul Krugman has written, in early 2020:

> *"Climate optimists have always hoped for a broad consensus in favor of measures to save our planet. The trouble with getting action on climate, the story went, was that it was hard to get people's attention: The climate was complex, while damage was too gradual and too invisible. In addition, the big dangers lay too far into the future. But surely once enough people had been informed about the dangers, once*

the global warming became sufficiently overwhelming, climate action would cease to be a partisan issue."[64]

Today's Question

In the 17th century, Blaise Pascal (1623 – 1662), the French mathematician, philosopher, and late in his short life theologian, was asked whether he believed in the existence of God. His answer was in the affirmative. For Pascal, the consequences he saw in *not* believing in God and being wrong about that would have been too much to bear.

In the early decades of the 21st century, the question we should be asking is what actions, if any, should be taken now to cope with global warming and the predictable consequences of climate change. Some – thankfully, a decreasing number – would argue we should just stand by until *all* scientists can come to full agreement. That wait-and-see approach would have been too much for Pascal to bear.

If actions to confront climate change are taken and no severely adverse outcomes in the climate follow, it would never be known with certainty whether those actions were really necessary. On the other hand, if no actions are taken and then some or all of the adverse consequences that climate scientists have predicted actually appear, it would never be known with certainty whether the actions not taken might have prevented those consequences. Certainty is not a choice, either way. We can hope that most will commit themselves to remedial actions before too much of our accustomed lifestyles are lost, but science is telling us that our climate is already changing and time is already wasting. Will the world's population, and in particular its political leaders, linger too long?

Chapter Eight
ARRIVAL

Flight 1230 touched down at the Beijing airport on schedule at 5:30 PM., local time, February 17. After a brief delay for traffic on the tarmac, the plane parked at its assigned gate and the passengers, weary after the long flight, collected their "belongings" and made their way off the plane and into the terminal. Those who lived in China were glad to be back, but those who were from other countries, especially those in China for the first time, were uneasy about the challenges they might confront in a foreign land.

The early chapters of this book dealt with five fundamentals of everyday life: numbers, measurements, calendars, clocks, and temperature. Now we are confronting the new kid on the block – global climate change. To deal with this new challenge, what have we learned from humankind's experiences, past and ongoing, from these earlier five fundamental changes in human life on this planet?

Numbers

It has taken over 30,000 years for the many cultures around the globe to come together with a common system of numbers, and even now different cultures are using different words and different symbols to represent the same number. The digit "3" means three, but people on the streets in Germany, Buenos Aires, and Beijing, for examples, all have their own words for "3" and those in some cultures also use their own symbols. But most importantly, in the 21st century we do have a common global understanding of quantities even if we still do not utter the same words or write the same symbols for equal quantities. For now, at least, our solution for these differences in expression when we encounter them is translation.

Our shared understanding of quantities can be attributed primarily to the worldwide growth of interchanges among cultures around the globe. The primary example comes from the spread of Arabic numbers, principally by Middle Eastern and North African cultures between about 700 CE to 1500 CE, to areas where they were settling in Spain. In what now seems like an extremely slow process, Arabic numbers first replaced traditional Roman numbers in the scientific minds of Europe and then even more gradually came to be accepted in the minds of clerics and European populations at-large. From there, Arabic numbers went with European settlers to the Americas, China and other distant destinations around the globe.

The contrast with climate change is apparent. Climate change disturbs balances in our natural world and, for many, forces traumatic changes in what had been settled life patterns. And for some, climate change even raises the question how they might be able to remake their lives in unfamiliar locations on our globe. Adopting a new system of numbers would not have been easy, but at least it was not forced by a change in the behavior of nature. In stark contrast, climate change does not wait. It is a natural force of its own.

Measurements

The movement towards a universal system of measurements, which is to say the metric system, while already favored by then current-day scientists, gained solid footing in the final decade of the 18th century when it became a stepchild of the French Revolution. One of the expressed goals of the revolutionists was to put all of France on a common system of both measuring and expressing units of weights and distances. Even before metric units had been officially defined, their usage was commanded in 1795 by the then-acting French legislative body. Those in charge of the effort had the foresight to invite other nations to join in preparing the final report about the new system. That report, which included the names and values of the units making up the new metric system, was released in 1799.

Over the course of the 19th century other nations in Europe, some motivated in part by Napoleon's conquering armies, adopted the metric system, as did many in South America. But Britain took its own path in 1824 by adopting its system of imperial units, which lasted within that nation and those under its control for about 150 years, until the mid-1970s. Then, as Britain was preparing to join the EU and, to that end, officially converting to the metric system, a range of issues emerged within a population reluctant to abandon the traditional imperial system. The switchover took many years and even now, well into the 21st century, a pint is still a pint in a British pub.

The switchover to the metric system has yet to occur in the United States, as far as the general population is concerned. Congress has passed the buck, so-to-speak, by approving the metric system but doing next-to-nothing to require its usage. Instead, the nation is enduring under what amounts to a "pseudo-traditional" system of measurements. It is "pseudo" because the traditional system of measurements in the United States (e.g., the foot and the pound), a cousin of the British imperial system, was maintained in name but not in content by the so-called Mendenhall Order, a published decision in 1893 by the then Superintendent of Weights and Measures, Thomas Corwin Mendenhall. Mendenhall declared that the international meter and the international kilogram would be the fundamental standards of length and weight in the United States. Since that action, when we now speak of a "foot" we really mean a fraction of a meter and when we speak of a "pound" we really mean a fraction of a kilogram. The general population in the United States has left itself locked into its traditional terminology while unwittingly referring in actuality to metric units, but its scientists have joined other scientists around the globe by expressly using the metric system.

China has a history of metrology which roughly resembles the United States experience. Various systems of measurement traditional to the various cultures within China had prevailed for centuries, until early in the 20th century, when the Chinese empress in 1908 took a step towards reform by sending a delegation to Paris to consult with the BIPM about conversion to the metric system. But political turmoil interrupted the effort, the empress was removed and the Republic of China was formed, only itself to be replaced in 1949 by the Communist party. In contrast with other western politics and lifestyles which were rejected by the Communists, the metric system was deemed to be a critical component of Chinese advancement. Due to ongoing political strife in China the system was given little attention until the 1980s, after Chairman Mao's death and the stabilization of political life. As in the United States, in the early decades of the 21st century China is living with two systems of measurement, the traditional system, which itself varies among regions and social classes, and the metric system, which has its footing in both the scientific community and those who have regular contacts with the West.

So our world is in the process of converting to the metric system, with the primary laggards being the general populations in both the United States and China. Those two nations, the two leading powers on the globe, are staggering forward towards widespread usage of the metric system. With ever growing trade and other interchanges, it is not unreasonable to assume the day will come when the globe in its entirety will be using metric.

The same assurance cannot be given, however, for humankind's dealing with climate change. Both the United States and China are major emitters of coal smoke into the globe's atmosphere, while other nations around the globe are continuing to emit as well. Meanwhile, scientists cannot predict when either the atmosphere or the oceans might become so overloaded with fossil fuel emissions and other contaminants that life as we know it will be irretrievably lost. The time clock is ticking against our capacity to maintain our lifestyle, much less to improve it.

Calendars and Clocks

It has taken more than five centuries for humankind worldwide to bring itself to a common method of keeping track of time for civic purposes, even if it does not do so for religious and other purposes. The bedrocks of this worldwide cohesion are the regularity, for all essential purposes, of the Earth's rotation around the Sun (the measure of one year) coupled with the regularity, for all essential purposes, of the Earth's rotation around its own axis (the measure of one day). Humankind has constructed on these two bedrocks its own measurements of months, weeks, hours, minutes and seconds, all of which are arbitrary concepts which now have been universally accepted. Using the combination of these time measurements, both the bedrocks and the constructed ones, humans around the globe have the capacity to coordinate their lives as they have become increasingly interdependent.

The many cultures now inhabiting the globe are being told by the large majority of climate scientists that the global climate is almost certainly in a state of change, due in large part to past decades of fossil fuel usage. Moreover, they are being told that the change seems to be accelerating, that there may well be points of no return, even some which might have already been crossed, and that the time to slow the change, or hopefully to reverse it, is unpredictable but almost certainly growing shorter. While the coordination of our calendars and clocks worldwide was set by the human desire to make life more compatible among many cultures with ever increasing interdependence, the direction and pace of climate change is being set by humankind's emissions of pollutants into the Earth's atmosphere.

Temperature

Temperature stands out among the five topics addressed in the earlier chapters of this book because it is an actual component of climate. Surprisingly, temperature was not a subject that grabbed our ancestors' attention until late in 17th century. Until then it was simply thought to be another phenomenon of nature, like rain or snow.

Temperature signals the commotion of particles, and consequently the presence of heat, within any substance having our attention. We use thermometers to measure temperature. The development of thermometers for general usage did not take hold until late in the

17th or early in the 18th century, relatively late in our scheme of things. The Age of Enlightenment brought a surge of interest in how things work, along with an active pursuit of ways to make life better. Across Europe a number of inventive persons dedicated themselves to improving and making thermometers, motivated in different degrees by curiosity, market demand, and profit.

Among the many possibilities, three types of thermometers survived the competition. Fahrenheit thermometers became a widely-accepted device in the 18th century. Soon thereafter, Celsius thermometers, using a metric scale, became a new standard and over the next two centuries displaced Fahrenheit in many places (a big exception is the United States, where Fahrenheit still prevails). During the 19th century, Lord Kelvin, a Scottish scientist who studied heat and methods of its measurement, continued the development of the Celsius thermometer. In 1954 the BIPM made temperature one of its basic units and in 1967 chose the word "kelvin" to designate one unit of heat. Now, in the 21st century, research is ongoing, primarily in laboratories, to meet the demands for ever more accurate measurements of heat, up and down an increasing range of temperatures.

Most of the global population has by now adopted the Celsius scale for measuring, reporting and talking about temperature, but in the United States the population is still holding on to the Fahrenheit scale, while its scientists, like other scientists around the globe, have converted to Celsius and kelvin measurements.

Temperature measurement can accommodate a lack of uniformity across national borders, as awkward and unnecessary as it may be, but climate change pays no heed to national borders. The climate is a global phenomenon and all global inhabitants need to treat it as such. The Paris Agreement of 2015, signed by 197 nations, was a promising first step towards the uniformity that is needed, but the steps that must now follow to achieve the goals of that agreement will require resolve by both political leaders and inhabitants in those nations. Time will tell.

Finding New Footing

First and foremost, accepting change in our everyday lives can be tough. It disrupts what is comfortably entrenched. It demands new thinking. It often requires outlays of money, sometimes large amounts. For these reasons and others, change often comes begrudgingly and, when viewed in hindsight, ever so slowly.

Second, in many cases change is accepted only when making it will bring advantages that outweigh sticking with the traditional. Conformity and convenience are glues holding us to what we already know. For one example, the United States Congress has had the power under the Constitution for more than 200 years to force a nationwide adoption of the metric system, but it has made only small gestures in that direction. The traditional U.S. system of units (yards, feet, inches, pounds, etc.) is very much alive and well within the nation's general population.

Third, in some cases change is accepted only when the traditional is no longer allowed. That is seen in the backstory about measurements. French revolutionaries late in the 18th century enacted laws to switch the French nation into the metric system, but 50 years later that still had not occurred in many places. Additional laws were enacted in the middle of the 19th century to prohibit traditional units, thereby forcing metric into everyday life across France. A similar experience was seen later in Britain. After that nation in the 1970s had begun to adopt the metric system, in anticipation of joining the EU, it took about two decades of bargaining with EU authorities so that the British people could, for example, continue to enjoy the "pint" at their local pubs.

Change does occur, however, but in hindsight with what in most cases seems like agonizing slowness. Ancient habitants, beginning with cave-dwellers and continuing for thousands of years afterwards, saw no need to learn more than what they needed to live on their own land and grow their own food. As other populations with different pursuits became more significant in their lives, the different populations found a need to communicate as trade among them was growing. Over time the lives of different groups became increasingly interdependent and that growth has both fostered and flowed out of that interdependence.

Much of this book has focused on the past, telling the story about changes that most humans around the globe have accepted because their lives are becoming more closely interconnected. But changes in the planet's climate, which have come into focus only since the middle of the 20th century, is a story still taking shape. It is much too early to draw any lessons about how we humans are dealing with it. More than that, climate change is an especially tough nut because, unlike other changes over centuries past, it affects all populations around the globe and, consequently, must be addressed by actions around the globe. Each sovereign nation can be expected to put priority on its own interests, while achieving cooperation among about 200 nations is an unprecedented international challenge. Only time will tell whether the risks that most climate scientists have forecasted will be addressed by enough sovereign nations, with enough commitment and within sufficient time, to make climate change a manageable problem.

Change is fundamentally a bottoms-up process. And with continuing advancements in communication and transportation, the diverse cultures still found around the globe can be expected to continue to bring themselves more closely together by accepting change in how to manage our climate. That process in many cases may well stretch across lifetimes. The optimistic viewpoint is that the human instinct to make life better – indeed, the instinct to preserve life – will ultimately prevail, as it has in other matters. If so, people in diverse cultures around the globe will increasingly understand that yes – we are all one humanity.

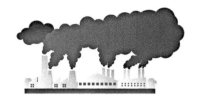

Appendix
SCIENCE – AN INTERNATIONAL ENTERPRISE

The International Bureau of Weights and Measures (in French, the *Bureau International des Poids et Mesures*, commonly known as the BIPM) is a key player in making science an international enterprise. Located in Sevres, France, a suburb of Paris, it is the international center for the science of measuring (metrology). The BIPM's work is the first domino in the lineup of new discoveries. Its work is fundamental to the progress of science around the globe.

Established in 1875 by France, the United States and other nations, the mission of the BIPM, stated most simply, is to set international standards of measurement. The BIPM is overseen by the General Conference on Weights and Measures (CGPM), an international body made up of delegates of the nations that sponsor its work.

The BIPM's standards of measurement consist of seven base units plus several so-called derived units, all of which are based upon the seven base units. The so-called SI units of the BIPM are expressed in the metric system, to the extent metric units are relevant. "SI" is the French acronym for *systém international d'unités*, or in English the international system of units. In the community of scientists, "SI" signals a BIPM base unit.

Having been launched as the international body to fix and promote uniform standards of measurement, the BIPM from time to time has refined several of those standards by expressing them not by reference to a physical object but, instead, by describing them in terms of a phenomenon found in nature. For example, in 1960 the BIPM changed its definition of the meter by setting aside the physical prototype stored in the French National Archives in Paris, which had been the standard for more than a century, and in its stead describing the meter to be the length of the path traveled by a beam of light in a tiny fraction of one second, within a vacuum. (The BIPM made a further refinement in 1983, but retained the length of a light path as the standard.)

The final step has been taken to separate BIPM's international standards of measurement from physical objects. Since the late 19th century, the kilogram has been defined by the weight of a casting in BIPM's custody outside Paris. In November 2018, a new definition, effective in May 2019, was approved by the CGPM, the international body that oversees the BIPM. The newly defined kilogram is now derived from a calculation based on Planck's constant, which can be thought of as the smallest amount of energy possible – a quantum theory concept. (This change came concurrently with changes in the definitions of the ampere, the kelvin and the mole, none of which, unlike the kilogram, involved a physical object.) The push for basing the BIPM's standards upon natural phenomena rather than prototypes has rested upon the inherent instability of physical objects, coupled with a desire to enable suitably equipped laboratories around the globe to reproduce its standards.

The BIPM's work reaches measurements of weight, length, volume and temperature and the tracking of time. It also extends to light, electricity, force and other topics that are of critical importance to scientists and, with little public awareness, to all of humankind. The BIPM arrived on stage about 300 years after the now commonly used Gregorian calendar first appeared. The calendar has not been a subject of its scrutiny, except to the extent that the BIPM occasionally adds, or potentially could subtract, a second of time to our clocks to keep them more accurately synchronized with the Earth's orbit around the Sun (see Chapter Five – Clocks). The gain or loss of a second is only a tiny speck on a calendar, but for many purposes a critically important speck.

Many of the BIPM's pronouncements catch little attention outside the realm of science, mostly because they involve a scale of accuracy which has little or no impact on everyday measurements. A refinement of the way to measure lengths which affects only digits many places to the right of a decimal point has no direct impact on the size of a shoe or a dining table. That same refinement, however, could well affect many of the investigations in scientific laboratories. The search for ever greater accuracy drives much of science. A pause to think about the intricacies of the multitude of tiny electrical circuits woven together inside a smartphone should engender some appreciation.

The demand for ever greater accuracy in science is never-ending. Those who are not immersed in science are not likely to comprehend much of what happens in the BIPM laboratories on the outskirts of Paris or in the other laboratories around the globe that collaborate with the BIPM. Even so, when the BIPM first sets a new international standard of measurement, or as more frequently happens when it redefines an existing standard, science uses that pronouncement and eventually it is reflected worldwide in things we all use in everyday life.

It is useful to think of the BIPM as the hub at the center of a wheel, with a collection of national laboratories around the rim. Other laboratories in universities and businesses, of course, make the wheel even larger.

NIST

In the United States, the National Institute of Science and Technology (NIST), an agency within the Department of Commerce, functions in close collaboration with the BIPM. Its two principal laboratories are at its headquarters in Gaithersburg, Maryland, outside Washington, D.C., and in Boulder, Colorado. Its staff numbers about 3,400 employees plus about 2,400 associates from academia, industry and other government agencies.

Among its extensive and ever-expanding list of projects, the NIST strives to promote a uniformity of measurements across the United States and to engender trust among all who rely upon uniformity. To that end, it works with the National Conference on Weights and Measures (NCWM), a body of state and local regulatory officials, on matters such as methods of measurement, allowable tolerances and labeling.

The NIST, established in 1902, was named the National Bureau of Standards until 1988. The name change reflects the expansion of the NIST's responsibilities, and that expansion, in turn, reflects the exponential growth of scientific pursuits during the 20th century, within the United States for sure but around the globe as well. The NIST, largely out of public view, plays a major role, side-by-side with industry, academia, other national laboratories and the BIPM, in the ongoing effort to push scientific frontiers forward.

Other National Laboratories

In Britain, the National Physical Laboratory (NPL) serves a function similar to that of the NIST. Other national laboratories similar to the NIST and the NPL are found in France (separate from the BIPM), Germany, China and several other countries. They all share information through channels maintained by the BIPM, and they all help the BIPM to fulfill its own international mission.

Science feeds upon itself and, for the most part, leaps across sovereign borders. Fortunately, scientists are not encumbered in the way that politicians and diplomats so routinely are. Scientifically, the BIPM is front and center on the international stage. National laboratories have a critically important and supportive role.

ABBREVIATIONS

BIPM – International Bureau of Weights and Measures (in French, the *Bureau international des poids et mesures*). This is the international organization based in Sevres, France, a suburb of Paris, that performs basic research and coordinates the research performed in national laboratories of its member nations, all for the purpose of fixing global standards of measuring units.

CGPM – General Conference on Weights and Measures. This is an international body made up of delegates of the nations that sponsor the work of the BIPM. It must approve changes in measuring units recommended by the BIPM.

COP – Conference of the Parties. This refers to the meetings concerning the global climate which are held from time to time at the call of the nations which are parties to the UNFCCC.

NIM – This is the Chinese counterpart of the NIST.

NIST – National Institute of Science and Technology. This is an agency within the United States Department of Commerce which performs basic research and performs tests for United States organizations. The NIST functions in collaboration with the BIPM and other national laboratories.

Abbreviations

NPL – National Physical Laboratory. This is the British counterpart of the NIST.

SI – French acronym for *systém international d'unités*, or in English the international system of units. In the community of scientists, SI signals a BIPM base unit.

TAI – International Atomic Time. This is a time determined by the BIPM, based upon readings from multiple atomic clocks around the globe. When TAI differs from UTC by 0.9 seconds, BIPM adjusts UTC to maintain coordination between the two time measurements.

UNFCCC – United Nations Framework Conference on Climate Change. This is a treaty adopted in 1992 by over one hundred nations at a conference in Rio de Janeiro.

UTC – Coordinated Universal Time. This is a time standard approved by the BIPM. It derives from the change adopted by CGPM in 1967 from using astronomical measurements to track time to using atomic clocks.

NOTES

Preface

1. Alexander Gerst, as quoted in "No Borders in Space: An Astronaut's Perspective of the Israeli-Palestinian Conflict," *The Higher Learning*, July 27, 2014, accessed on November 5, 2016, http://thehigherlearning.com/2014/07/27/no-borders-in-space-an-astronauts-perspective-of-the-israeli-palestinian-conflict/. At the time, Gerst was a German astronaut aboard the international space station.

Chapter Two – Numbers

1. Chinese numbers and words in the following table have been taken from Georges Ifrah, *The Universal History of Numbers*, 263 - 273 (New York, NY: John Wiley & Sons, Inc. 2000).
2. Ibid., 15.
3. Ibid., 5.
4. Edward P. Burger, *Zero to Infinity: A History of Numbers* (course guidebook for lecture series, The Great Courses: 2007) 5.
5. Ifrah, *The Universal History of Numbers*, 68 - 70.
6. John McLeish, *Number – The history of numbers and how they*

Notes

shape our lives, 56 - 58 (New York, NY: Ballantine Books, 1992).

7. Ifrah, *The Universal History of Numbers*, 288 – 296.
8. Ibid., 205, 590.
9. Ibid., 206, 590.
10. McLeish, *Number*, 65 – 71.
11. Ibid., 66
12. Ifrah, *The Universal History of Numbers*, 577
13. Ibid., 578 – 579.
14. Karl Menninger, *Number Words and Number Symbols – A Cultural History of Numbers*, (New York, NY: The MIT Press, original English edition 1969, republished as a Dover edition 1992), 461 – 462.
15. Denise Schmidt-Besserat, with illustrations by Michael Hay, *The History of Counting*, (New York, NY: HarperCollins Books 1999), 12; Joshua Project, "Gilyak, Nivkh in Russia," accessed February 28, 2017, https://joshuaproject.net/people_groups/11894/RS.
16. Ifrah, *The Universal History of Numbers*, 532 – 539.
17. Ibid., 532 – 533, 555.
18. Ibid., 534 – 539.
19. Ibid., 263 – 273; McLeish, *Number*, closely resembling 55; Meninger, *Number Words*, 457.
20. Ibid., 158.
21. Ibid., 303 -304.
22. Ibid., 437 – 439; recent carbon dating of an ancient Indian text indicates that a symbol closely resembling what we now call a zero was used as a place holder as early as the 3rd century, see "History of zero pushed back 500 years by ancient text," *New Scientist*, September 14, 2017, accessed September 16,

2017, https://newscientist.com/article/2147450-history-of-zero-pushed-back-500-years-by-ancient-Indian-text/.
23. McLeish, *Number*, 11, 82; Sarah Gibbens, "Ancient Text Reveals New Clues to the Origin of Zero," September 16, 2017, *National Geographic,* http://news.nationalgeographic.com/2017/09/origin-zero-bakhshali-manuscript-video-spd/.
24. Burger, *Zero to Infinity*, 15; Ifrah., *The Universal History of Numbers*, 399; McLeish, *Number*, 140.

Chapter Three – Measurements

1. U.S. Metric Association, "Unit Mixups – Gimli Glider: Boeing 767 emergency landing," accessed March 6, 2017, http://www.us-metric.org/unit-mixups/#gimli; Wikipedia contributors, "Gimli Glider," *Wikipedia, The Free Encyclopedia*, https://en.wikipedia.org/w/index.php?title=Gimli_Glider&oldid=768596633 (accessed March 6, 2017).
2. U.S. Metric Association, "Unit Mixups – Loss of Mars Climate Orbitor," accessed March 6, 2017, http://www.us-metric.org/unit-mixups/#mco.
3. Robert P. Crease, *World in the Balance – The Historic Quest for an Absolute System of Measurement* (New York, NY: W.W. Norton & Company, Inc., 2011), 259.
4. Ibid., 18, 73.
5. Ibid, 73.
6. Crease, *World in the Balance*, 19, 43 – 45. Withold Kula, *Measures and Men*, trans. R. Szeter (Princeton, NJ: Princeton University Press, 1986), 284.
7. Wikipedia contributors, "Stone (unit)," *Wikipedia, The Free Encyclopedia*, https://en.wikipedia.org/w/index.php?title=Stone_(unit)&oldid=768996200 (accessed March 13, 2017).

Notes

8. Kula, *Measures and Men*, ch. 4.
9. Crease, *World in the Balance*, 69 – 88.
10. Ibid., 88.
11. See U.S. Metric Association, "Metrication in other countries – Advance of metric usage in the world," accessed March 7, 2017, http://www.us-metric.org/metrication-in-other-countries/#chart.
12. Wikipedia contributors, "Metrication in the United Kingdom," *Wikipedia, The Free Encyclopedia*, https://en.wikipedia.org/wiki/Metrication_in_the_United_Kingdom (accessed March 4, 2019). This source provides a review of metrication in Britain from the early 18th century to the 21st century and has been a source for much of the information that follows in the text about that topic.
13. Wikipedia contributors, "Metrication Board," *Wikipedia, The Free Encyclopedia*, https://en.wikipedia.org/w/index.php?title=Metrication_Board&oldid=725920857 (accessed March 4, 2019).
14. Ibid., see list of industries and dates of conversion under the heading "Activities of the Board."
15. Ibid., see the heading "Education".
16. Wikipedia contributors, "European units of measurement directives," *Wikipedia, The Free Encyclopedia*, https://en.wikipedia.org/w/index.php?title=European_units_of_measurement_directives&oldid=880321289 (accessed March 5, 2019).
17. Ibid.
18. UK Metric Association, "Historical perspectives by the last Director of the UK Metrication Board," accessed March 7, 2017, http://www.metric.org.uk/articles/jhumble.
19. Ibid.

20. Wikipedia contributors, "Metrication in the United Kingdom," heading "Current usage."
21. Crease, *World in the Balance*, 109
22. Ibid., 109 – 114.
23. Lewis V. Judson, Weights and Measures Standards of the United States, a brief history, NBS Special Publication, originally issued October 1963, updated March 1976, 3, accessed March 8, 2017, https://www.nist.gov/sites/default/files/documents/pml/wmd/pubs/2010/12/16/sp-447-2.pdf
24. Judson, Weights and Measures Standards of the United States, 4-5; Crease, *World in the Balance*, 117 – 121.
25. Judson, Weights and Measures Standards of the United States, 3
26. Ibid., 6-8.
27. Ibid., 8
28. Ibid.
29. Crease, *World in the Balance*, 131.
30. Wikipedia contributors, "Metric Act of 1866," *Wikipedia, The Free Encyclopedia*, https://en.wikipedia.org/w/index.php?title=Metric_Act_of_1866&oldid=764238669 (accessed March 9, 2017).
31. Crease, *World in the Balance*, 134 – 138.
32. Ibid., 253; Wikipedia contributors, "Metric Conversion Act," *Wikipedia, The Free Encyclopedia*, https://en.wikipedia.org/w/index.php?title=Metric_Conversion_Act&oldid=742659111 (accessed March 9, 2017).
33. Presidential Documents, Exec. Order No. 12770, 56 Fed. Reg. 145 (July 25, 1991).
34. U.S. Metric Association, "Unit Mixups – Winning long jump record lost." accessed March 7, 2017, http://www.us-metric.org/unit-mixups/#longjump.

35. Crease, *World in the Balance*, 42.
36. Ibid., 144 – 146.
37. Kula, *Measures and Men*, 284 – 286.
38. Crease, *World in the Balance*, 218.
39. Ibid.
40. China Expats, "Chinese Weights and Measures," accessed March 8, 2017, http://www.china-expats.com/WeightsMeasures.htm; Wikipedia contributors, "Chinese units of measurement," *Wikipedia, The Free Encyclopedia*, https://en.wikipedia.org/w/index.php?title=Chinese_units_of_measurement&oldid=766272610 (accessed March 8, 2017).
41. CUSP, "National Institute of Metrology, China," accessed July 5, 2017, http://en.nim.ac.cn/.

Chapter Four – Calendars

1. G. J. Whitrow, *Time in History – Views of Time from Prehistory to the Present Day* (New York, NY: Barnes & Noble, Inc. 1988), 83.
2. Ibid., 116 – 118.
3. Ibid.
4. Ibid.
5. Ibid.
6. Ibid., 119
7. Ibid., 120.
8. Ibid., 74.
9. William Safire, "B.C./A.D. or B.C.E./C.E.?," *New York Times Magazine*, August 17, 1997, accessed March 25, 2017, http://www.nytimes.com/1997/08/17/magazine/bc-ad-or-bce-ce.html.
10. Whitrow, *Time in History*, 31 – 32.
11. Ibid., 32.

12. Ibid., 26.
13. Ibid., 54.
14. On the Jewish calendar, years are numbered by counting from what followers of that religion believe to have been the date of creation. See Judaism 101, "Jewish Calendar," accessed March 17, 2017, http://www.jewfaq.org/calendar.htmf.
15. Whitrow, *Time in History*, 44 – 45.
16. Ibid., 66 -67.
17. New World Encyclopedia, "Jesuit China missions," accessed March 25, 2017, http://www.newworldencyclopedia.org/entry/Jesuit_China_missions.
18. Timeanddate.com, "The Chinese Calendar," accessed March 25, 2017, https://www.timeanddate.com/calendar/about-chinese.html.
19. Ibid.; for a more complete description of the traditional Chinese calendar, see ChinaKnowledge.de, "An Encyclopedia on Chinese History, Literature and Art – The Chinese Calendar," accessed March 25, 2017, http://www.chinaknowledge.de/History/Terms/calendar.html.
20. The Mayan calendars are explained in Smithsonian National Museum of the American Indian, "Living Maya Time – The Calendar System," accessed March 17, 2017, https://maya.nmai.si.edu/calendar/calendar-system.
21. Whitrow, *Time in History*, 66.
22. Ibid., 66 – 67.
23. Oxford Dictionaries, "How did the months get their names?," accessed March 27, 2017, http://blog.oxforddictionaries.com/2016/01/months-names/; Old Farmer's Almanac, "Origin of Month Names," accessed March 27, 2017, http://www.almanac.com/content/origin-month-names.

Notes

24. Christian Bible Reference Site, "What Does the Bible Say About the Sabbath?," accessed March 19, 2017, http://www.christianbiblereference.org/faq_sabbath.htm; Whitrow, *Time in History*, 70.
25. Oxford Dictionaries, "The names of days of the week in English and other languages," accessed March 27, 2017, http://blog.oxforddictionaries.com/2012/08/just-plutonic/;Old Farmer's Almanac, "Origin of Day Names," accessed March 27, 2017, http://www.almanac.com/content/origin-day-names.
26. Ian R. Bartky, *One Time Fits All* (Stanford, CA, Stanford University Press 2007), 9 – 10.
27. Ibid., 10.
28. Ibid., 11.
29. Ibid., 21 – 22.
30. Ibid., 23 – 26.
31. Wikipedia contributors, "Anglo-French Conference on Time-keeping at Sea," *Wikipedia, The Free Encyclopedia*, https://en.wikipedia.org/w/index.php?title=Anglo-French_Conference_on_Time-keeping_at_Sea&oldid=752899182 (accessed April 5, 2017).

Chapter Five – Clocks

1. G. J. Whitrow, *Time in History – Views of Time from Prehistory to the Present Day* (New York, NY: Barnes & Noble Books 2004), 17.
2. We distinguish the two parts of a day with the terms "AM" (ante meridiem) and "PM" (post meridiem). These designations derive from the Latin word *meridies*, meaning midday. Wikipedia contributors, "Meridian (geography)," *Wikipedia, The Free Encyclopedia*, https://en.wikipedia.org/w/index.

php?title=Meridian_(geography)&oldid=809035605 (accessed November 8, 2017).
3. Robert Lamb, "Water-powered Clocks Through the Ages," accessed April 1, 2017, http://science.howstuffworks.com/environmental/green-tech/sustainable/water-powered-clock1.htm.
4. Wikipedia contributors, "Water clock," *Wikipedia, The Free Encyclopedia*, https://en.wikipedia.org/w/index.php?title=Water_clock&oldid=772952683 (accessed April 1, 2017); Daniel Mintz, "Timekeeping in the Ancient World: Water-clocks," accessed April 1, 2017, http://www-groups.dcs.st-and.ac.uk/history/HistTopics/Water_clocks.html.
5. NIST, "A Walk Through Time – Early Clocks," accessed April 1, 2017, https://www.nist.gov/pml/walk-through-time-early-clocks; Gerhard Dorhn-van Rossum, *History of the Hour – Clocks and Modern Temporal Order,* trans. Thomas Dunlap (Chicago: University of Chicago Press, 1996), 84 – 88; Wikipedia contributors, "Water clock," Wikipedia contributors, "Water clock," *Wikipedia, The Free Encyclopedia*, https://en.wikipedia.org/w/index.php?title=Special:CiteThisPage&page=Water_clock&id=774042423.
6. Whitrow, *Time in History*, 78.
7. William Harris, "How Astrolabes Work," accessed May 11, 2017, http://electronics.howstuffworks.com/gadgets/clocks-watches/astrolabe.htm.
8. Dorhn-van Rossum, *History of the Hour*, 46.
9. Jeremy Norman, "Huygens Invents the Pendulum Clock, Increasing Accuracy Sixty Fold," accessed April 2, 2017, http://www.historyofinformation.com/expanded.php?id=3506.
10. Ibid.
11. Whitrow, *Time in History*, 166.

Notes

12. Wikipedia contributors, "Vasco da Gama," *Wikipedia, The Free Encyclopedia*, https://en.wikipedia.org/w/index.php?title=Vasco_da_Gama&oldid=768934473 (accessed April 2, 2017).
13. Whitrow, *Time in History*, 140 – 141.
14. David S. Landes, *Revolution in Time −Clocks and the Making of the Modern World*, rev. ed. (Cambridge, MA: Harvard University Press 2000), 116.
15. Whitrow, *Time in History*, 140 – 141.
16. For a more complete account of Harrison's experience, see Landes, *Revolution in Time*, ch. 9.
17. Whitrow, *Time in History*, 146.
18. BIPM, "SI Brochure: The International System of Units (SI) [8th edition, 2006; updated 2014]," accessed April 3, 2017, http://www.bipm.org/en/publications/si-brochure/second.html.
19. NIST, "NIST Researcher David Wineland Wins 2012 Nobel Prize in Physics," accessed April 11, 2017, https://www.nist.gov/news-events/news/2012/10/nist-researcher-david-wineland-wins-2012-nobel-prize-physics; Economist, "Technology Quarterly: Quantum Devices – Here, there and everywhere," *The Economist*, March 11, 2017, accessed April 11, 2017, http://www.economist.com/technology-quarterly/2017-03-09/quantum-devices; also see Adam Minter, "Lutetium and ytterbium are vying to become the beating heart of science and the global economy," Bloomberg Businessweek, August 27, 2019, accessed September 1, 2019, https:// https://www.bloomberg.com/news/features/2019-08-28/the-world-s-most-precise-clock-may-run-on-lutetium-or-ytterbium.
20. NIST, "NIST Time Frequently Asked Questions (FAQ) - Why is UTC used as the acronym for Coordinated

Universal Time instead of CUT?,": accessed April 3, 2017, https://www.nist.gov/pml/time-and-frequency-division/nist-time-frequently-asked-questions-faq.
21. BIPM, "International Atomic Time," accessed April 4, 2017, http://www.bipm.org/en/bipm-services/timescales/tai.html; NIST, "Web Clock Frequently Asked Questions (FAQ) – What leap seconds are, how they are implemented and their future," Accessed April 4, 2017. https://www.nist.gov/pml/time-and-frequency-division/popular-links/web-clock-faq.
22. Find UTC for your time zone on the internet at http://time.gov.
23. Whitrow, *Time in History*, 158 – 159.
24. Ibid., 161.
25. Ibid., 165.
26. Ian R. Bartky, *One Time Fits All* (Stanford, CA: Stanford University Press 2007), 48.
27. Ian R. Bartky, *Selling the True Time – Nineteenth-Century Timekeeping in America* (Stanford, CA: Stanford University Press 2000), ch. 2.
28. Bartky, *One Time Fits All*, 70 – 72.
29. Ibid., 120 – 127; Vanessa Ogle, "A Briefer History of Time – How the World Adopted A Uniform Conception of Time," *Foreign Affairs*, October 12, 2015, accessed April 5, 2017, https://www.foreignaffairs.com/articles/2015-10-12/briefer-history-time.
30. Bartky, *One Time Fits All*, 127 – 134.
31. Wikipedia contributors, "Time in China," *Wikipedia, The Free Encyclopedia*, https://en.wikipedia.org/w/index.php?title=Time_in_China&oldid=773359095 (accessed April 5, 2017).
32. Matt Schiavenza, "China Only Has One time Zone – and That's

a Problem," *The Atlantic*, November 5, 2013, accessed April 5, 2017, https://www.theatlantic.com/china/archive/2013/11/china-only-has-one-time-zone-and-thats-a-problem/281136/.

33. Far West China, "Xinjiang Time – A Tale of Two Time Zones," accessed April 5, 2017, https://www.farwestchina.com/2015/02/xinjiang-time-a-tale-of-two-time-zones.html. North Korea presents a different source of confusion, having reset its own time zone in 2015 to run 30 minutes behind that of nearby South Korea and Japan. Accessed August 14, 2017, http://www.cnn.com/2015/08/07/asia/north-korea-time-zone/index.html.

34. Whitrow, *Time in History*, 165; Landes, *Revolution in Time*, 304.

35. Arthur E. Zimmerman, "The First Wireless Time Signals to Ships at Sea," accessed April 5, 2017, http://www.antiquewireless.org/uploads/1/6/1/2/16129770/50-the_first_wireless_time_signals_to_ships_at_sea.pdf.

36. When GPS is used to fix a position, at least four satellites are working in unison. The precision in their coordination directly affects the precision of the position they indicate on the GPS device. An overview of navigation methods can be found at: National Geospatial-Intelligence Agency, "Introduction to Marine Navigation," accessed April 16, 2017, http://msi.nga.mil/MSISiteContent/StaticFiles/NAV_PUBS/APN/Chapt-01.pdf.

37. Timeanddate, "Daylight Saving Time Statistics," accessed February 26, 2019, https://www.timeanddate.com/time/dst/statistics.html.

38. Wikipedia contributors, "Daylight saving time by country," *Wikipedia, The Free Encyclopedia*, https://en.wikipedia.

org/w/index.php?title=Daylight_saving_time_by_ country&oldid=772420132 (accessed March 28, 2017); "Call time on putting the clocks forward, Europe tells Brussels," *Financial Times*, August 30, 2018.

39. Matt Schiavenza, "China Only Has One Time Zone – and That's a Problem," *Atlantic*, accessed February 28, 2019, https://www.theatlantic.com/china/archive/2013/11/china-only-has-one-time-zone-and-thats-a-problem281136/.
40. Bartky, *One Time Fits All*, 163 – 173.
41. Ibid., 173 – 182.
42. Royal Museums Greenwich, "British Summer Time (BST) and the Daylight Saving Time plan," accessed April 6, 2017, http://www.rmg.co.uk/discover/explore/british-summer-time-and-daylight-saving.
43. Bartky, *One Time Fits All*, 186.
44. Ibid., 189 – 195.
45. Ibid., 198 – 199.
46. Wikipedia contributors, "Daylight saving time in the United States," *Wikipedia, The Free Encyclopedia*, https://en.wikipedia.org/w/index.php?title=Daylight_saving_time_in_the_United_States&oldid=771850532 (accessed April 7, 2017).

Chapter Six – Temperature

1. Online Etymology Dictionary, "thermometer," accessed March 12, 2017, http://www.etymonline.com/index.php?allowed_in_frame=0&search=thermometer
2. In thermodynamics, scientists use the word "entropy" to communicate about the degree of disorder within a closed system. Treating a porch as a closed system, if all items on the porch were at the same temperature (same internal commotion of

Notes

particles) there would, in theory, be maximum disorder and, in thermodynamic theory, maximum entropy. For more explanation, see "Entropy as Time's Arrow," accessed March 12, 2017, http://hyperphysics.phy-astr.gsu.edu/hbase/Therm/entrop.html#e3.

3. The designation SI is explained in the Appendix.
4. W. E. Knowles Middleton, *A History of the Thermometer and Its Use in Meteorology*, (Baltimore MD: The Johns Hopkins Press 1966) 4-5.
5. Ibid., 3-4.
6. Ibid. 62.
7. Ibid., 5-8.
8. Ibid., 11-12.
9. Ibid., 28.
10. Ibid., 68-69
11. Ibid., 79.
12. Ibid., 98.
13. Ibid., 98-101.
14. Ibid., 101-105.
15. Ibid., 91; U.S. Metric Association, "Metric system temperature (kelvin and degree Celsius)," accessed March 15, 2017, http://www.us-metric.org/metric-system-temperature-kelvin-and-degree-celsius/.
16. Herbert Arthur Klein, *The Science of Measurement – A Historical Survey*, Dover ed. 1988 (New York, NY: Simon & Schuster 1974), 313.
17. Wikipedia contributors, "William Thomson, 1st Baron Kelvin," *Wikipedia, The Free Encyclopedia*, https://en.wikipedia.org/w/index.php?title=William_Thomson,_1st_Baron_Kelvin&oldid=769260803(accessed March 15, 2017).

18. B.W. Mangum, "Report of the 17th Session of the Consultative Committee on Thermometry – Special Report on the International Temperature Scale of 1990," *Journal of Research of the National Institute of Standards and Technology 95, no.1*, accessed March 12, 2017, http://nvlpubs.nist.gov/nistpubs/jres/095/jresv95n1p69_A1b.pdf.
19. Michael Bicay et al., "Infrared Radiation: More Than Our Eyes Can See," accessed March 15, 2017, http://coolcosmos.ipac.caltech.edu/resources/paper_products/print_publication_pdf/IRUback.pdf.

Chapter Seven – Climate

1. Ed Lorenz, address to the American Association for the Advancement of Science, Washington, D.C., December 29, 1979, as quoted in Spencer R. Weart, *The Discovery of Global Warming* (Cambridge MA: Harvard University Press, rev ed. 2008), 114; "Understanding the full scope of human impacts on the climate requires a global focus because of the interconnected nature of the climate system," D.J. Wuebbles et al., *Climate Science Special Report: Fourth National Climate Assessment, Volume I*, U.S. Global Change Research Program (2017), accessed November 8, 2017, https://science2017globalchange.gov/, 23; also see *Volume II* of the cited report, November 23, 2018, accessed February 11, 2019, https://nca2018.globalchange.gov/chapter/front-matter-about#chfront-matter-about-1.
2. Gerst, see quote in Preface.
3. Jeffrey Bennett, *A Global Warming Primer: The Science, the Consequences, and the Solutions* (Boulder CO: Big Kid Science, 2016), 51; Spencer Weart, "Carbon Dioxide Greenhouse Effect --The Computer Models Vindicated

Notes

(1990s-2000s)," *The Discovery of Global Warming* (online version, January 2017), accessed January 26, 2017, http://history.aip.org/climate/co2.htm, 21; John Cook, et al., "Quantifying the consensus on anthropological global warming in the scientific literature," (IOP Publishing 2013), accessed January 26, 2017, http://iopscience.iop.org/article/10.1088/1748-9326/8/2/024024/pdf.

4. William J. Ripple, Christopher Wolf, Thomas M. Newsome, Mauro Galetti, Mohammed Alamgir, Eileen Crist, Mahmoud I. Mahmoud, William F. Laurence and 15,364 scientist signatories from 184 countries, *World Scientists' Warning to Humanity: A Second Notice*, accessed November 19, 2017, https://academic.oup.com/bioscience; Wuebbles et al., *Climate Science Special Report*, 11, 13, 14, 17, and 31. Using a geological time scale, rather than our time scale for humankind, our time on this planet has been compared by geologist Robert M. Hazen to a walk across the United States, from New York City to San Francisco, with each step westward representing a century back in time. Recorded history ends before we can reach the New York sidewalk, and when we reach the Pacific Ocean we will have covered only ten percent of the Earth's history (a time span that includes ice ages, dinosaurs, and other major climate changes). That comparison leads to the conclusion that "[i]n this century alone [our first step], a time scale so laughably brief as to effectively not exist to geologists, we could send the planet back to a climate system not seen for many millions of years." Peter Brannen, "Rambling Through Time," *New York Times*, January 27, 2018, accessed February 8, 2018, https://www.nytimes.com/2018/01/27/opinion/rambling-through-time.html; IPCC, Summary for Policymakers,

accessed January 11, 2020, https://ippc.ch/site/assets/uploads/sites/2/2019/05/SR15_SPMversion_report_LR.pdf.

5. On many issues, such as medical treatments and solar eclipses, most of us trust the work and predictions of scientists. On climate change, however, we are slow to react to the predictions that most climate scientists have been making, in part because they require large scale, collective actions as well as changes in our daily lives, but also because they threaten economic interests. Justin Gillis, "Should You Trust Climate Science? Maybe the Eclipse Is a Clue," *New York Times*, August 20, 2017, accessed December 6, 2017, https://www.nytimes.com/2017/08/18/climate/should-you-trust-climate-science-maybe-the-eclipse-is-a-clue.html.

6. Ripple et al., *World Scientists' Warning to Humanity: A Second Notice*; Wuebbles et al., *Climate Science Special Report*, 32. The New York Times has published a three-part report about an ongoing scientific investigation into the behavior of the Antarctica's ice sheet. In a worst-case scenario, disintegration of the ice sheet has become unstoppable and could cause the ocean to rise six feet or more worldwide by the end of the 21st century. Scientists emphasize the need for more data to predict future behavior. NYT, "Miles of Ice Collapsing into the Sea," Somini Sengupta, "U.N. Chief Warns of Dangerous Tipping Point on Climate Change," *New York Times,* September 10, 2018, https://www.nytimes.com/2018/09/10/climate-united-nations-climate-change.html.

7. James Rodger Fleming, *Historical Perspectives on Climate Change*, chaps. 1 through 4 (New York, NY: Oxford University Press, 1998).

8. John Mason, *The History of Climate Science*, "In the beginning

...," posted April 7, 2013, accessed January 27, 2017, www.skepticalscience.com/history-climate-science.html.
9. Ibid., "Tyndall and heat-trapping gases."
10. Ibid., "Carbon dioxide: Arrhenius makes a discovery."
11. Spencer Weart, *The Discovery of Global Warming* rev. ed. (Cambridge, MA: Harvard University Press, 2008), 10. (Explanatory note: Weart is the author of both the book cited here and the supplemented online version cited in note 3 above. The two versions use the same title. The book you are reading uses both as sources.)
12. Weart, *The Discovery of Global Warming* (online version), "The Carbon Dioxide Greenhouse Effect -- Callendar's Advocacy."
13. Weart, *The Discovery of Global Warming* (book), 31 – 33, 38 – 62.
14. Carbon Dioxide Information Analysis Center, accessed December 28, 2016, http://cdiac.ornl.gov/trends/co2/vostok.html and http://cdiac.ornl.gov/trends/co2/ice_core_co2.html.
15. Weart, *The Discovery of Global Warming* (online version), "The Carbon Dioxide Greenhouse Effect -- Time Lags and Feedbacks (1900s),"; "Milankovitch Cycles and Glaciation," accessed January 27, 2017, http://www.indiana.edu/~geol105/images/gaia_chapter_4/milankovitch.htm.
16. Mason, *The History of Climate Science*, "Ice cores and ancient air."
17. Ibid., "Solving the Faint Young Sun Paradox."
18. Spencer Weart, "Other Greenhouse Gases – Struggling toward Policies," *The Discovery of Global Warming* (online version), accessed January 29, 2017, http://history.aip.org/climate/other-gas.htm

19. Henry Fountain, "Alaska's Permafrost Is Thawing," August 23, 2017, https://www.nytimes.com/interactive/2017/08/23/climate/alaska-permafrost-thawing.html; Wuebbles et al., *Climate Science Special Report*, 23 and 29; for more information about methane, see "Climate change – The methane mystery," *Economist*, April 28, 2018, 71, and John Schwartz and Brad Plumer, "the Natural Gas Industry Has a Leak Problem," *New York Times*, June 21, 2018.
20. Wuebbles et al., *Climate Science Special Report*, 31.
21. NASA Earth Observatory, "Features -- Ocean and Climate," accessed January 29, 2017, http://earthobservatory.nasa.gov/Features/OceanClimate/ocean-atmos_phys.php; Wuebbles et al., *Climate Science Special Report*, 25.
22. Caitlyn Kennedy, "Why did the Earth's surface temperature stop rising in the past decade?," accessed April 17, 2017, https://www.climate.gov/news-features/climate-qa/why-did-earth%E2%80%99s-surface-temperature-stop-rising-past-decade.
23. Joseph Romm, *Climate Change – What Everyone Needs to Know* (New York, NY: Oxford University Press 2016), 6; Wuebbles et al., *Climate Science Special Report*, 25.
24. See Kendra Pierre-Louis, "2019 Was a Record Year for Ocean Temperatures, Data Show," *New York Times*, January 13, 2020, accessed January 14, 2020, https://nyti.ms/384oHDO.
25. Scientists attribute the Great Ocean Conveyor Belt to "thermohaline circulation," which means that both temperature and salinity differentials are primary causes. National Ocean Service, "Currents – Thermohaline Circulation," accessed May 16, 2017, http://oceanservice.noaa.gov/education/tutorial_currents/05conveyor1.html; NASA, "The Thermohaline

Notes

Circulation – The Great Ocean Conveyor Belt," accessed May 16, 2017, https://pmm.nasa.gov/education/videos/thermohaline-circulation-great-ocean-conveyor-belt; Wikipedia contributors, "Thermohaline circulation," *Wikipedia, The Free Encyclopedia*, https://en.wikipedia.org/w/index.php?title=Thermohaline_circulation&oldid=780443686 (accessed May 19, 2017). A fuller explanation can be found in Stefan Rahmstorf, "Ocean Currents and Climate Change," accessed May 16, 2017, http://www.pik-potsdam.de/~stefan/Lectures/ocean_currents.html. Of special interest to navigators in the Atlantic is the warm Gulf Stream which flows northeastward toward the British Isles. Benjamin Franklin is credited for his recognition of the Gulf Stream, late in the 18th century, and his creation of the original map showing that it is a part of what has since come to be called the Great Ocean Conveyor Belt. See John Elliott Pillsbury, United States Coast and Geodetic Survey, Report of 1890, Appendix 10, Ch II, accessed May 19, 2017, https://docs.lib.noaa.gov/rescue/oceanheritage/Gc296g9p541891.pdf.

26. Ibid. The absorbed CO_2, however, is not uniformly distributed throughout the ocean. "How the oceans store CO_2 is critical for understanding the global carbon cycle," Pacific Marine Environmental Laboratory, accessed January 29, 2017, http://www.pmel.noaa.gov/co2/story/Ocean+Carbon+Storage.

27. United Nations Framework Convention on Climate Change (UNFCCC), "Fast facts & figures: On the Bali Road Map and Cancun Agreement," accessed January 29, 2017, http://unfccc.int/essential_background/basic_facts_figures/items/6246.php.

28. Wuebbles et al., *Climate Science Special Report*, 11; Weart, *The Discovery of Global Warming* (book), 28; Mason, *The*

History of Climate Science, "1980s: The Carbon-Cycle - Earth's thermostat"; historical data about CO_2 emissions since 1900 can be found in William Nordhaus, *Climate Casino – Risk, Uncertainty, and Economics for a Warming World* (New Haven, CT and London 2013); recent monthly data about CO_2 in the atmosphere is published online by the Earth System Research Laboratory and can be found at https://www.esrl.noaa.gov/gmd/ccgg/trends/.

29. Bennett, *A Global Warming Primer*, 18; a range of projections for future global warming and consequential effects on climate, based upon four scenarios for emissions of greenhouse gasses can be found in Wuebbles et al., *Climate Science Special Report*, ch. 4.

30. U.S. Energy Information Administration, accessed April 19, 2017, https://www.eia.gov/tools/faqs/faq.php?id=73&t=11.

31. U.S. Energy Information Administration, "Residential Energy Consumption Survey (RECS) – RECS data show decreased energy consumption per household," https://www.eia.gov/consumption/residential/reports/2009/consumption-down.php?src=%E2%80%B9%20Consumption%20%20%20%20%20%20Residential%20Energy%20Consumption%20Survey%20(RECS)-f4#fig-4.

32. Intergovernmental Panel on Climate Change, "Climate Change 2014 Synthesis Report, Fifth Assessment Report (2014)", accessed January 29, 2017, http://ar5-syr.ipcc.ch/; Nordhaus, *The Climate Casino*, chaps. 6 through 12; Romm, Climate Change, chaps. 2 and 3; Weart, *The Discovery of Global Warming* (online version) – Impacts of Climate Change; Ripple et al., *World Scientists' Warning to Humanity: A Second Notice*, figure 1 and page 2.

33. Union of Concerned Scientists Fact Sheet, "The Science Connecting Extreme Weather to Climate Change," June 2018, https:// ucsusa.org/sites/default/files/attach/2018/06/The-Science-Connecting-Extreme-Weather-to-Climate-Change.pdf.
34. Christopher Flavelle, "Climate Change Threatens the World's Food Supply, United Nation Warns," accessed August 18, 2019, https:// nytimes.com/2019/08/08/climate/climate-change-food-supply-html?searchResultPosition=1; also see Robinson Meyer, "This Land Is the Only Land There Is," accessed August 18, 2019, https://theatlantic.com/science/archive/2019/08/how-we-think-about-the-dire-new-ipcc-climate-report/595705/.
35. A summary of the study appears in Brandon Miller, "Satellite observations show sea levels rising, and climate change is accelerating it," accessed February 23, 2018, https://www.cnn.com/2018/02/12/world/sea-level-rise-accelerating/index.html; also see Chris Mooney, "Antarctica ice loss has tripled in a decade. If that continues, we are in serious trouble.," *Washington Post*, June 13, 2018; John Schwartz, "Greenland's Melting Ice Nears a 'Tipping Point,'" *New York Times*, January 21, 2019; Eric Rignot, Jeremie Mouginot et al, "Four decades of Antarctica Ice Sheet mass balance from 1979–2017," *Proceedings of the National Academy of Sciences*, January 22, 2019, accessed February 22, 2019, http://www.pnas.org/content/116/4/1095#abstract-2.
36. Hannah Devlin, "Receding glacier causes immense Canadian river to vanish in four days," *The Guardian*, April 17, 2017, accessed April 21, 2017, https://www.theguardian.com/science/2017/apr/17/receding-glacier-causes-immense-canadian-river-to-vanish-in-four-days-climate-change; John Schwartz, "Climate Change Reroutes a Yukon River in a Geological

Instant," *New York Times*, April 17, 2017, accessed April 21, 2017, https://www.nytimes.com/2017/04/17/science/climate-change-glacier-yukon-river.html?_r=0.

37. Rachel Carson, *Silent Spring*, ann. ed. (Boston, MA: Houghton Mifflin Harcourt, 2002).
38. Al Gore, *Earth in the Balance*, rev. ed. (Boston, MA: Houghton Mifflin Harcourt, 2000); see Nathaniel Rich, "Losing Earth: The Decade We Almost Stopped Climate Change," *New York Times Magazine*, August 1, 2018, accessed February 24, 2019, https://nytimes.com/interactive/2018/08/01/magazine/climate-change-losing-earth.html.
39. A sequel titled *"An Inconvenient Sequel: Truth to Power"* was released in 2017.
40. Pope Francis, "Laudito Si – On Care for Our Common Home" (Encyclical Letter, 2015), accessed February 18, 2017, http://w2.vatican.va/content/francesco/en/encyclicals/documents/papa-francesco_20150524_enciclica-laudato-si.html: published summary of key excerpts, Sarah Pulliam Bailey, "10 key excerpts from Pope Francis's encyclical on the environment," *Washington Post*, June 18, 2015, accessed on February 18, 2017, https://www.washingtonpost.com/news/acts-of-faith/wp/2015/06/18/10-key-excerpts-from-pope-franciss-encyclical-on-the-environment/?tid=a_inl&utm_term=.9b5108eef726.
40. An analogy has been made between the public's reaction, prior to the millennial change, concerning the so-called Y2K computer threat and the public's attitude about the threat of climate change.

If you want to prompt expensive, collective global action, you need to tell people the absolute worst that could happen. We humans do not stir at the merely slightly uncomfortable.

Farhad Manjoo, "How Y2K Offers a Lesson for Fighting Climate Change," *New York Times*, July 23, 2017; a report has been issued by the Environmental Data and Governance Initiative, a non-profit organization, about the removal of climate change information from online sites of federal agencies during the Trump administration, "Changing the Digital Climate – How Climate Change Web content is Being Censored Under the Trump Administration," accessed January 12, 2018, https://envirodatagov.org/wp-content/uploads/2018/01/Part-3-Changing-the-Digital-Climate.pdf.

41. Wikipedia, "Environmental policy in China," accessed January 29, 2017, https://en.wikipedia.org/wiki/Environmental_policy_in_China.
42. Shimon Peres (1923 – 2016), the former President and Prime Minister of Israel, in a 1995 interview said,

Science knows no borders, technology has no flag, information has no passport. The new challenges transcend the old notion of boundaries.

Nathan Gardels, editor-in-chief, "Weekend Roundup: With Cyber, There Are No Front Lines In War Or Peace," posted April 1, 2017, *The World Post*, accessed April 21, 2017, http://www.huffingtonpost.com/entry/weekend-roundup-163_us_58de6a82e4b0e6ac709451b3?6seljhw52rbxzuxr&section=us_world.

44. UNFCCC, text of United Nations Framework Convention on Climate Change 1992, Article 2, page 9, accessed January 29, 2017, http://unfccc.int/files/essential_background/background_publications_htmlpdf/application/pdf/conveng.pdf.
45. Wikipedia, "Kyoto Protocol," accessed February 6, 2017, https://en.wikipedia.org/wiki/Kyoto_Protocol#Non-ratification_by_the_USA; UNFCCC, "Kyoto Protocol," accessed February 6, 2017, http://unfccc.int/kyoto_protocol/items/2830.php.
46. John Vidal, Allegra Stratton and Suzanne Goldenberg, "Low targets, goals dropped: Copenhagen ends in failure," *The Guardian*, December 18, 2009, accessed January 29, 2017, https://www.theguardian.com/environment/2009/dec/18/copenhagen-deal.
47. IPCC, Climate Change 2014 Synthesis Report Summary for Policymakers, accessed January 9, 2018, see note 31 above, updated in Summary for Policymakers, October 6, 2018, World Meteorological Association, accessed February 14, 2019, https://www.ipcc.ch/sr15/chapter/summary-for-policy-makers.
48. UNFCCC, "Lima Climate Change Conference – December 2014," accessed January 30, 2017, http://unfccc.int/meetings/lima_dec_2014/meeting/8141.php; "Lima Call for Climate Action Puts World on Track to Paris 2015," accessed January 30, 2017, http://newsroom.unfccc.int/lima/lima-call-for-climate-action-puts-world-on-track-to-paris-2015/.
49. Text of Paris Agreement, see note 37 above; Mercator Research Institute on Global Commons and Climate Change maintains an online time clock showing the time that remains until the temperature limits of the Paris Agreement would be exceeded, see https://www.mcc-berlin.net/forschung/co2-budget.html.

50. Wuebbles et al., *Climate Science Special Report*, 32.
51. NASA Earth Observatory, "World of Change -- Global Temperatures," accessed January 30, 2017, https://earthobservatory.nasa.gov/Features/WorldOfChange/decadaltemp.php; another source asserts a 1.0°C increase since 1901, see Wuebbles et al., *Climate Science Special Report,* Volume I, 10, 17 (Volume II was released on November 26, 2018, https://nca2018.globalchange.gov).
52. Text of speech, June 1, 2017, https://www.whitehouse.gov/the-press-office/2017/06/01/statement-president-trump-paris-climate-accord.
53. Associated Press, "US Tells UN It Is Pulling Out of Paris Climate Deal," *New York Times*, November 4, 2019, https://www.nytimes.com/apoline/2019/11/04/business/bc-us-trump-climate1st-ld-writethru.html; Paris Agreement, Article 28, Sections 1 and 2; UNFCCC, Paris Agreement Status of Ratification, accessed June 3, 2017, http://unfccc.int/paris_agreement/items/9444.php.
54. Wuebbles et al., *Climate Science Special Report*, 12.
55. Brad Plumer, "Climate Negotiators Reach an Overtime Deal to Keep Paris Pact Alive," *New York Times*, December 15, 2018, https://www.nyt.com/2018/12/15/climate/cop24-katowice-climate-summit.html.
56. Nathan Gardels, editor-in-chief, "Weekend Roundup: Economic hardship and nationalism are gutting climate action," posted December 8, 2018, *The World Post*, accessed December 8, 2018, http://www.theworldpost.com (this short article highlights a few of the tradeoffs encountered, in various localities around the globe, between jobs, present energy demand and climate protection); also see Neil Irwin, "Climate Change's

Giant Impact on the Economy: 4 Key Issues," *New York Times*, January 17, 2019, http://nytimes.com/2019/01/17/upshot/how-to-think-about-the-costs-of-climate-change.html.

57. An organization of "non-federal" actors, calling itself America's Pledge, has come forward to speak on behalf of those within the United States who are committed to helping the country fulfill its obligations under the Paris Agreement. In that organization's Phase 1 Report, released November 2017, it claims to have mobilized more than half of the United States economy, while asserting nevertheless that federal engagement is critical. The organization's website (accessed November 13, 2017) is https://www.americaspledgeonclimate.com/. These self-appointed representatives of the United States, in the absence of official representatives, are participating in ongoing meetings with official representatives of other nations which are signatories of the Paris Agreement. "New life for the Paris deal," *Economist*, December 16, 2017. Additionally, young adults have initiated a nationwide movement calling for action on climate change (Alexandra Yoon-Hendricks, "Meet the Teenagers Leading a Climate Change Movement," *New York Times*, July 21, 2018).

57. The need for negative emissions to meet the 2°C (3.6°F) goal was made apparent in 2014 in the IPCC's release of its Climate Change 2014 Synthesis Report, cited in note 31 above; the difficulties that lie in achieving negative emissions have been highlighted in Kevin Anderson, Glen Peters et al, "The trouble with negative emissions," *Science*, October 14, 2016, 182; efforts to develop technologies for achieving negative emissions are discussed in Abby Rabinowitz and Amanda Simson, "The Dirty Little Secret of the World's Plan to Avert Climate

Disaster," accessed January 10, 2018, https://www.wired.com/story/the-dirty-secret-of-the-worlds-plan-to-avert-climate-disaster/, and in both the editorial "What they don't tell you," at 11, and the accompanying article "Sucking up carbon," at 20, *Economist*, November 18 – 24, 2017. Methods being considered to reduce emissions include, among others, (a) design of buildings (see "Construction and carbon emissions – Home truths about climate change," *Economist*, January 5, 2019), (b) use of recyclable containers ("The Milkman Model: Big Brand Names Try Reusable Containers," *New York Times*, January 24, 2019), (c) afforestation and reforestation (Anderson, Peters et al, "The trouble with negative emissions," 182), (d) addition of carbon uptake catalysts to soils and oceans (Anderson, Peters et al, "The trouble with negative emissions," 182) and (e) direct-air-capture (Jon Gertner, "The Tiny Swiss Company That Thinks It Can Help Stop Climate Change," *New York Times Magazine*, February 12, 2019).

58. Pew Research Center in 2015 conducted a survey in 40 nations, involving 45,435 respondents, to measure public concern about climate change. The survey did not attempt to measure individual commitment. Survey results were published in November 2015, shortly before the international Paris climate conference. Results varied among regions and along political, gender and religious lines. The results can be found at "Global Concern about Climate Change, Broad Support for Limiting Emissions," November 5, 2015, accessed May 2, 2017, http://www.pewglobal.org/2015/11/05/global-concern-about-climate-change-broad-support-for-limiting-emissions/. The Center in 2016 conducted another survey within the United States, involving 1,534 adults, to measure attitudes about climate

change. The results showed important divisions between political groups about the fact and causes of climate change, the need to confront it, and the level of trust to be placed in climate scientists. This 2016 survey also addressed the distinction between concern and individual commitment. While 75 percent of the respondents said they were concerned about helping the environment, only 20 percent described themselves to be making an effort in their everyday lives to protect the environment all the time. The 2016 survey results can be found at "The Politics of Climate," October 4, 2016, accessed May 2, 2017, http://www.pewinternet.org/2016/10/04/the-politics-of-climate/; also see "In the line of fire," *Economist*, August 4, 2018. The delay in confronting climate change has now generated consideration among climate scientists and other experts about pursing adaptation, alongside mitigation, to inevitable changes in our global climate. Editors, "Adapt or Mitigate? Both," *Scientific American*, December 2019, p. 9, accessed November 28, 2019, https://www.scientificamerican.com/article/adapt-or-mitigate-both/. Also see J. Christensen and A. Olhoff, "Lessons from a decade of emissions gap assessments," United Nations Environment Programme,, Nairobi, https://www.unenvironment.org/resources/emissions-gap-report-10-year-summary; Brady Dennis, "In bleak report, U.N. says drastic action is only way to avoid worst effects of climate change," https://washingtonpost.com/climate-environment/2019/11/26/bleak-report-un-says-drastic-action-is-only-way-avoid-worst-impacts-climate-change/.

59. UN Climate Action Summit 2019, p. 1, accessed January 14,2020, https://un.org/en/climatechange/un-climate-summit-2019.shtml.

Notes

60. International Panel on Climate Control, Summary for Policymakers, https://ipcc.ch/site /assets/uploads/sites/2/2019/05/SR15_SPM_version_report_LR.pdf.
61. See note 59 above; also see Report of the Secretary-General on the Climate Action Summit and The Way Forward in 2020, https://https://un.org/en/climatechange/assets/pdf/cas_report_11_dec.pdf.
62. Report of the Secretary-General cited in note 61 above.
63. Somini Sengupta, "U.N. Climate Talks End With Few Commitments and a 'Lost' Opportunity," New York Times December 15, 2019, accessed January 17, 2019, https:// nytimes.com/2019/12/15/climate/cop25-un-climate-talks-madrid.html?searchResultPosition=3.
64. Paul Krugman, "Australia shows us the road to hell," accessed January 13, 2020, https://nytimes.com/2020/01/09/opinion/australia-fires-html?searchResultPosition=4; also see *Economist*, "Sustainable investing – Green Giant – The world's largest asset manager promises to be more climate-friendly," January 18, 2020, 72.

Appendix

1. The BIPM homepage is http://www.bipm.org/en/about-us/.
2. Information about the units of measurement defined by the BIPM can be found from BIPM's homepage (see note 1 above) by clicking on Measurement Units. For more detail, see NIST Special Publication 330 (2008 Ed.), Barry N. Taylor and Ambler Thompson, eds., accessed March 11, 2017, https://www.nist.gov/sites/default/files/documents/pml/div684/fcdc/sp330-2.pdf.
3. The NIST homepage is https://www.nist.gov/.

BIBLIOGRAPHY

Several of the sources listed here are accompanied by brief descriptions prepared by the author. These are included to help readers decide which sources might be most responsive to their interests.

Preface

Higher Learning. "No Borders in Space: An Astronaut's Perspective of the Israeli-Palestinian Conflict." July 27, 2014. Accessed November. 5, 2016. http://thehigherlearning.com/2014/07/27/no-borders-in-space-an-astronauts-perspective-of-the-israeli-palestinian-conflict/.

Overbye, Dennis. "Jacob Bekenstein, Physicist Who Revolutionized Theory of Black Holes, Dies at 68." *International New York Times*, August 21, 2015.

Chapter One – Departure

BIPM. Accessed February 21, 2017. http://www.bipm.org/en/about-us/, plus companion web pages.

National Physical Laboratory (NPL). Accessed February 27, 2017. http://www.npl.co.uk/, plus companion web pages.

Bibliography

NIST. Accessed February 21, 2017. https://www.nist.gov/, plus companion web pages.

Chapter Two – Numbers

Burger, Edward B. *"Zero to Infinity: A History of Numbers."* Course guidebook for recorded series of 24-lectures. The Great Courses 2007.

Ifrah, Georges. *The Universal History of Numbers – From Prehistory to the Invention of the Computer.* New York, NY: John Wiley & Sons, Inc. 2000.

> Ifrah (1947 -) was once a school teacher who was both embarrassed and inspired by a class of students who posed questions he could not answer, indeed ones that had never before come to his mind. He left his teaching job and, while sponsoring himself with odd jobs, spent years traveling the globe in search of answers. This book, a selection of the Book-of-the Month Club, is a compendium of his discoveries. In its 600, double-columned, over-sized pages, it gives an encyclopedic treatment to the history of numbers, covering prehistoric counting methods and moving through the centuries to our modern Hindu- Arabic system. The book sweeps in cultures, past and present, around the globe. Numerous illustrations accompany the text. Originally written in French, the book is now in English translation. The index can be helpful to those seeking specific information. Ifrah's work in its entirety can be absorbed by those with the time and commitment to do so. It is not difficult to read, just lengthy, tending in places to

wander a bit off the track.

Joshua Project. "Gilyak, Nivkh in Russia." Accessed February 28, 2017. https://joshuaproject.net/people_groups/11894/RS

McLeish, John. *Number – The history of numbers and how they shape our lives*. New York, NY: Ballantine Books 1992. (This book has also been published in paperback with the title *The Story of Numbers – How Mathematics Has Shaped Civilization*.)

> McLeisch was a professor at the University of Victoria, British Columbia. This book, about 250 pages, was a selection of the Book-of-the Month Club. It follows an historic path, from preliterate to ancient to contemporary societies. McLeish states in his introduction that one purpose is "to show how different number systems arose in different societies, and how each helped to shape the society which devised it." He laments the negative influence that early Greek and later anti-scientific Christian attitudes during the Dark Ages had upon the development of numbers in Europe, contrasting that with the progress that was meanwhile being made in the hands of Indian and Arabic mathematicians. This covers much of the same historical material that Menninger's book details (see below), but it is a shorter and easier read.

Menninger, Karl. *Number Words and Number Symbols – A Cultural History of Numbers.* The MIT Press, original English edition 1969, republished as a Dover edition 1992.

Bibliography

Menninger (1898 – 1963) was a German mathematician, not the American psychiatrist of the same name who wrote several books in his own field. In his preface, Menninger asserts that the book's purpose is to "trace [the] interweaving of numbers and human life." The bulk of attention is given to the history of numbers in various cultures, their appearance, usage and eventual replacement. Some reviewers, without questioning the substance, have found the writing somewhat dense and stilted; others express only respect for the in-depth treatment. The book contains numerous illustrations. The final section, while shorter than its particular topic might deserve, gives a good explanation of number systems in China, Japan and Korea.

Merzbach, Uta C. and Carl B. Boyer, *A History of Mathematics*, 3rd ed. Chaps. 1 – 4 and 9 – 12. Hoboken, NJ: John Wiley & Sons, Inc., 2011.

Morley, Iain and Colin Renfrew, eds. *The Archaeology of Measurement – Comprehending Heaven, Earth and Time in Ancient Societies*. New York, NY: Cambridge University Press, 2010.

The editors hold positions at Oxford and Cambridge universities and other research institutions. The book is a collection of 20 scholarly essays about measuring in ancient times and the relationship of measuring to the development of societies. Some of the essays are accompanied by illustrations.

Schmidt-Besserat, Denise, with illustrations by Michael Hay. *The History of Counting*. New York, NY: HarperCollins Books 1999.

In this colorfully illustrated book, the author, an archaeologist, gives a brief survey of various counting systems used in different cultures over the centuries of discoverable human history. She offers a somewhat closer look at the benefits of the Hindu-Arabic number system. This is a quick read for anyone who might want to compare our number system with those that have not survived in the flow of time and scientific advancement.

Chapter Three – Measurements

China Expats. "Chinese Weights and Measures." Accessed March 8, 2017. http://www.china-expats.com/WeightsMeasures.htm.

China Highlights. "Units of Measurement in China." Accessed March 8, 2017. http://www.chinahighlights.com/travelguide/guidebook/china-units-of-measurements.htm.

Crease, Robert P. *World in the Balance – The Historic quest for an Absolute System of Measurement.* New York, NY: W.W. Norton & Company, Inc., 2011.

CUSP. "National Institute of Metrology, China." Accessed March 8, 2017. http://cuspbj.com/En/CompanyNews/2015326213342666.html

Ganeri, Anita. *The Story of Weights and Measures.* New York, NY: Oxford University Press, Inc., 1996.

This is a 30-page, large-sized book with numerous illustrations in color. It is in the nature of a short encyclopedia about measuring systems during the known history of humankind.

Bibliography

Judson, Lewis V. *Weights and Measures Standards of the United States, a brief history*, NBS Special Publication, originally issued October 1963, updated March 1976. Accessed March 8, 2017. https://www.nist.gov/sites/default/files/documents/pml/wmd/pubs/2010/12/16/sp-447-2.pdf.

> Information about the author has not been found. This publication, available on the internet, is a well-written chronological history of the development of measuring systems in the United States from the time of early settlers to the 1970s.

Klein, Herbert Arthur. *The Science of Measurement – A Historical Survey*, Dover ed. 1988, chs. 1 through 9. New York, NY: Simon & Schuster 1974.

> Klein is the author of several books about metrology, the science of measurement. This book of more than 700 pages, republished in 1988, offers a detailed explanation of modern measuring systems coupled with summaries of their development over recent times. It is a very readable piece of work, backed by extensive research. As its page count indicates, the author had serious readers in mind.

Kula, Witold. *Measures and Men*, trans. R. Szreter. Princeton, NJ: Princeton University Press, 1986.

> Kula (1916 – 1988), while a professor at the University of Warsaw, was a founder and president of the International Economic History Association. In this book he has written about the transition from pre-modern methods of

measurement, typically reflecting local practices, to modern methods based upon widespread acceptance of quantitative standards. The book is recommended to those readers who have a serious curiosity about the evolution of modern measuring standards.

Marciano, John Bemelmans, *Whatever Happened to the Metric System – How America Kept Its Feet*. New York, NY: Bloomsbury USA 2014.

Marciano is the author and illustrator of various books, including some for children. His focus in this book is the history of the metric system from the French Revolution onward, with an emphasis on why the general population in the United States still clings to its customary units rather than embracing metric ones. The reader will encounter in this book, interwoven with its primary discussion of the metric system, much briefer discussions of time zones, daylight saving time, language uniformity, calendar reform and currency.

Morley, Iain and Colin Renfrew, eds. *The Archaeology of Measurement – Comprehending Heaven, Earth and Time in Ancient Societies*. New York, NY: Cambridge University Press, 2010.

A brief description of this book appears in this Bibliography under Chapter 2 - Numbers.

Presidential Documents, Exec. Order No. 12770. 56 Fed. Reg. 145. July 25, 1991.

Bibliography

Schwartz, David M., with pictures by Steven Kellogg. *Millions to Measure*. New York, NY: HarperCollins Publishers 2003.

> This is a large-sized, 42-page, highly illustrated book printed in color, written and illustrated to introduce young children to the concepts of measurement.

UK Metric Association. "Historical perspectives by the last Director of the UK Metrication Board." Accessed March 7, 2017. http://www.metric.org.uk/articles/jhumble.

U.S. Metric Association. "Metrication in other countries – Advance of metric usage in the world." Accessed March 7, 2017. http://www.us-metric.org/metrication-in-other-countries/#chart.

U.S. Metric Association. "Unit Mixups – Gimli Glider: Boeing 767 emergency landing." Accessed March 6, 2017. http://www.us-metric.org/unit-mixups/#gimli.

U.S. Metric Association. "Unit Mixups – Winning long hump record lost." Accessed March 7, 2017. http://www.us-metric.org/unit-mixups/#longjump.

Whitelaw, Ian. *a measure of all things – The Story of Man and Measurement*. New York, NY: St. Martin's Press 2007.

> Whitelaw is an editor and writer. Within the 157 pages of this book, which includes illustrations and portraits, he presents a survey of many units of measurement used in multiple settings and for different purposes, from atomic to cosmic. Brief historical perspectives are given for some, but not all.

Wikipedia contributors. "Chinese units of measurement." *Wikipedia, The Free Encyclopedia.* https://en.wikipedia.org/w/index.php?title=Chinese_units_of_measurement&oldid=766272610. Accessed March 8, 2017.

Wikipedia contributors. "European units of measurement directives." *Wikipedia, The Free Encyclopedia.* https://en.wikipedia.org/w/index.php?title=European_units_of_measurement_directives&oldid=756884823. Accessed March 7, 2017.

Wikipedia contributors. "Gimli Glider." *Wikipedia, The Free Encyclopedia.* https://en.wikipedia.org/w/index.php?title=Gimli_Glider&oldid=768596633. Accessed March 6, 2017.

Wikipedia contributors. "Metric Act of 1866." *Wikipedia, The Free Encyclopedia.* https://en.wikipedia.org/w/index.php?title=Metric_Act_of_1866&oldid=764238669. Accessed March 9, 2017.

Wikipedia contributors. "Metrication Board." *Wikipedia, The Free Encyclopedia.* https://en.wikipedia.org/w/index.php?title=Metrication_Board&oldid=725920857 Accessed March 7, 2017.

Wikipedia contributors. "Metrication in the United Kingdom." *Wikipedia, The Free Encyclopedia.* https://en.wikipedia.org/w/index.php?title=Metrication_in_the_United_Kingdom&oldid=769106647. Accessed March 7, 2017.

Wikipedia contributors. "Metric Conversion Act." *Wikipedia, The Free Encyclopedia.* https://en.wikipedia.org/w/index.php?title=Metric_Conversion_Act&oldid=742659111. Accessed March 9, 2017.

Bibliography

Chapter Four – Calendars

Bartky, Ian R. *One Time Fits All.* Stanford, CA, Stanford University Press 2007.

> Bartky (1924 – 2007) was trained as a physical chemist and spent most of his career at the NBS (later becoming the NSIT). He acquired an interest in timekeeping and time dissemination and in the 1970s was consulted by Congress about extending the nationwide period of daylight saving time. Bartky earned a reputation of being both a world expert on timekeeping and a careful scholar. His book, *One Time Fits All,* addresses not only the International Date Line, covered in this chapter, but also time zones and daylight saving time, covered in Chapter Five.

ChinaKnowledge.de. "An Encyclopedia on Chinese History, Literature and Art – The Chinese Calendar." Accessed March 25, 2017. http://www.chinaknowledge.de/History/Terms/calendar.html.

Christian Bible Reference Site. "What Does the Bible Say About the Sabbath?" Accessed March 19, 2017. http://www.christianbiblereference.org/faq_sabbath.htm.

Judaism 101. "Jewish Calendar." Accessed March 17, 2017. http://www.jewfaq.org/calendar.htm.

Maestro, Betsy, illustrated by Giulo Maestro. *The Story of Clocks and Calendars.* New York, NY: Harper Collins Publishers 1999.

Betsy Maestro (1944 -), the author, taught kindergarten and first grade for 11 years before turning her attention, in collaboration with her husband, an illustrator, to the preparation of more than 50 books for children. This book is one in that series. It is a full-page, colorfully illustrated treatment aimed primarily at young readers. In its 48 pages the book gives a short excursion into the history of both calendars and clocks.

Marciano, John Bemelmans. "The Great Calendar Debate," chap. 13/16 in *Whatever Happened to the Metric System? – How America Kept Its Feet.* New York, NY: Bloomsbury USA 2014.

A brief description of this book appears in this Bibliography under Chapter Two – Numbers.

New World Encyclopedia. "Jesuit China missions." Accessed March 25, 2017. http://www.newworldencyclopedia.org/entry/Jesuit_China_missions.

Old Farmer's Almanac. "Origin of Month Names." Accessed March 27, 2017. http://www.almanac.com/content/origin-month-names.

Oxford Dictionaries. "How did the months get their names?" Accessed March 27, 2017. http://blog.oxforddictionaries.com/2016/01/months-names/.

Philip, Alexander. *The Calendar: Its History, Structure And Improvement...* . Cambridge: Cambridge University Press 1921.

Alexander Philip (1858 – 1932) was a Scottish solicitor and a campaigner for reform of the Gregorian calendar. This book

Bibliography

is a reprint of his early 20th century writing giving a detailed history of calendars and assessing both the strengths and weaknesses he saw in the calendar decreed by Pope Gregory XIII late in the 16th century. His analysis will find primary interest in those who want a detailed understanding of the Gregorian calendar and the movement late in 19th century and early 20th century to reform it.

Safire, William. "B.C./A.D. or B.C.E./C.E.?," *New York Times Magazine*, August 17, 1997. Accessed March 25, 2017. http://www.nytimes.com/1997/08/17/magazine/bc-ad-or-bce-ce.html.

Smithsonian National Museum of the American Indian. "Living Maya Time – The Calendar System." Accessed March 17, 2017. https://maya.nmai.si.edu/calendar/calendar-system.

Timeanddate.com. "The Chinese Calendar. Accessed March 25, 2017. https://www.timeanddate.com/calendar/about-chinese.html.

Whitrow, G. J. *Time in History – Views of Time from Prehistory to the Present Day*. New York, NY: Barnes & Noble, Inc. 1988.

Whitrow (1912 – 2000) was a professor at the University of London and Senior Research Fellow of the Imperial College of Science and Technology, London. His primary interests were in mathematics, cosmology and the history of science. This book, in its highly readable 200 pages, deals both with the tracking of time in periods of years, months and days (the calendar) and with tracking it in periods of hours and minutes (the clock). The narrative reflects a philosophical perspective about the passage of time, while also addressing how humans historically have attempted to measure it.

Wigelsworth, Jeffrey R., *Science and Technology in Medieval European Life*, in the series daily life through history, Westport, CT: Greenwood Press 2006, ch. 8.

Chapter Five – Clocks

Bartky, Ian R. *Selling the True Time – Nineteenth-Century Timekeeping in America*. Stanford, CA: Stanford University Press 2000. Information about Bartky appears in this Bibliography under Chapter Four.

Dohrn-van Rossum, Gerhard. *History of the Hour – Clocks and Modern Temporal Orders*. Translated by Thomas Dunlap. Chicago, IL: University of Chicago Press 1996.

> Dohrn-van Rossum teaches medieval and early modern history at the University of Bielefeld in Germany and has been a visiting professor at the University of Chicago and the University of Zurich, Switzerland. This book gives a detailed history of the emergence of mechanical clocks in Europe from the late Middle Ages to the industrial revolution, giving attention to the transformation of humankind's time-consciousness with the transition from an agrarian to an industrial society.

Johnston, Andrew K., Roger D. Connor, Carlene E. Stephens, and Paul E. Ceruzzi, "Time and Place Connection" and "Navigating at Sea" in *Time and Navigation – The Untold Story of Getting from Here to There*. Washington, D.C.: Smithsonian Books 2015.

Bibliography

Landes, David S. *Revolution in Time − Clocks and the Making of the Modern World*, rev. ed. 2000. Cambridge, MA: Harvard University Press 2000.

Landes (1924 − 2013) was a professor of history and economics at Harvard who in the 1970s found that he had a consuming interest in watches. His pursuit of that interest coupled itself to his ongoing interest in cultural and economic development, and that coupling led to his writing of the first edition of *Revolution in Time*, published in 1983. The cited book is the revised edition, published in 2000. With a primary focus on Europe, this book in its 394 pages of text (plus extensive notes) offers a general history of time measurement and its contribution to modern civilization.

Wigelsworth, Jeffrey R., *Science and Technology in Medieval European Life*, in the series daily life through history, Westport, CT: Greenwood Press 2006, ch. 8.

Chapter Six – Temperature

Bicay, Michael, Michelle Thaller, Linda Hermans, Mike Bennett, Schuyler Van Dyk, and Richard Yessayian. "Infrared Radiation: More Than Our Eyes Can See." Accessed March 15, 2017. http://coolcosmos.ipac.caltech.edu/resources/paper_products/print_publication_pdf/IRUback.pdf

Bolton, Henry Carrington. *Evolution of the Thermometer*. San Bernardino, CA: The Perfect Library 2016.

Carrington (1843 – 1903) was a chemist and bibliographer who studied, worked and taught in both Europe and the United States during the last half of the 19th century. This is a short book (64 pages) offering historical details about thermometers.

Klein, Herbert Arthur. *The Science of Measurement – A Historical Survey,* Dover ed. 1988., chs. 26 through 29. New York, NY: Simon & Schuster 1974.

Klein and this book are briefly described under Chapter 3 – Measurements. Chapters 26 through 29 deal with thermal energy and its measurement.

Mangum, B.W. "Report of the 17th Session of the Consultative Committee on Thermometry – Special Report on the International Temperature Scale of 1990." *Journal of Research of the National Institute of Standards and Technology* 95, no.1. Accessed March 12, 2017. http://nvlpubs.nist.gov/nistpubs/jres/095/jresv95n1p69_A1b.pdf.

Middleton, W.E. Knowles. *A History of the Thermometer and Its Use in Meteorology.* Baltimore, MD: Johns Hopkins Press, 1966.

Middleton (1902 – 1998) spent most of his life in Canada, where as a physicist he held a range of positions in both academia and industry related to weather and instrumentation. He wrote numerous books and articles on these topics and received wide recognition for his work. The book cited here (paperback, 238 pages, easily readable) gives a good overview of the development of thermometers, including the efforts of numerous inventors involved in that effort.

Moldover, Michael R., Weston L. Tew and Howard W. Yoon. "Box 1: The international temperature scale of 1990." *Nature Physics* vol. 12, 7 – 11. Published online Jan. 7, 2016. Accessed August 12, 2016. http://www.nature.com/nphys/journal/v12/n1/box/nphys3618_BX1.html

U.S. Metric Association. "Metric system temperature (kelvin and degree Celsius)" Accessed March 15, 2017. http://www.us-metric.org/metric-system-temperature-kelvin-and-degree-celsius/.

Wikipedia contributors. "William Thomson, 1st Baron Kelvin." *Wikipedia, The Free Encyclopedia.* https://en.wikipedia.org/w/index.php?title=William_Thomson,_1st_Baron_Kelvin&oldid=769260803. Accessed March 15, 2017.

Chapter Seven – Climate

Bailey, Sarah Pulliam. "10 key excerpts from Pope Francis's encyclical on the environment." *Washington Post, June 18, 2015.* Accessed on February 18, 2017. https://www.washingtonpost.com/news/acts-of-faith/wp/2015/06/18/10-key-excerpts-from-pope-franciss-encyclical-on-the-environment/?tid=a_inl&utm_term=.9b5108eef726

Bennett, Jeffrey. A Global Warming Primer. Boulder, CO: Big Kid Science 2016.

Bennett (1932 -), educated as an astrophysicist, specializes in mathematics and science education. He is the lead author of college textbooks in a variety of scientific subjects and other books for the general public, including children. *A Global Warming Primer*, in its 102 illustrated pages, presents in question and answer format the basic evidence that global warming is occurring, answers to contentions made by skeptics, expected consequences of warming, and potential solutions. In a short read it gives a good overview of both the science and the politics.

Carbon Dioxide Information Analysis Center. Accessed December 28, 2016. http://cdiac.ornl.gov/trends/co2/vostok.html and http://cdiac.ornl.gov/trends/co2/ice_core_co2.html,

Carson, Rachel. *Silent Spring*, ann. ed. Boston, MA: Houghton Mifflin Harcourt 2002).

Fleming, James Rodger. *Historical Perspectives on Climate Change*. New York: Oxford University Press 1998.

Fleming (1949 -) is a professor at Colby College and research associate at the Smithsonian Institution. He has been a participant in various professional organizations related to weather and climate and has given testimony to Congressional committees. This book (paperback), in its 139 pages of text plus 50 pages of notes, gives historical details about weather and climate.

Gillis, Justin, illustrations by Jon Han. "Climate Change Is Complex. We've Got Answers to Your Questions." September 19, 2017. https://www.nytimes.com/interactive/2017/climate/what-is-climate-change.html?_r=0.

Bibliography

Gore. Al. *Earth in the Balance*, rev. ed. Boston, MA: Houghton Mifflin Harcourt, 2000.

Intergovernmental Panel on Climate Change. Fifth Assessment Report 2014. Accessed January 30, 2017. http://www.ipcc.ch/publications_and_data/publications_and_data_reports.shtml.

As mentioned in the text, the IPCC is an organization established jointly in 1988 by the United Nations Environment Program and the World Meteorological Organization to provide information about climate change and its impacts. It has 195 member nations. Scientists from around the world contribute to its work. This 2014 report is its most recent. It carries the following as a headline conclusion: "Continued emission of greenhouse gases will cause further warming and long-lasting changes in all components of the climate system, increasing the likelihood of severe, pervasive and irreversible impacts for people and ecosystems. Limiting climate change would require substantial and sustained reductions in greenhouse gas emissions which, together with adaptation, can limit climate change risks."

NASA. "Climate Kids." Accessed January 29, 2017. http://climatekids.nasa.gov/ocean/.

NASA Earth Observatory. "Features -- Ocean and Climate." Accessed January 29, 2017. http://earthobservatory.nasa.gov/Features/OceanClimate/ocean-atmos_phys.php.

Mason, John. *The History of Climate Science*. Posted April 7, 2013. Accessed January 27, 2017. http://www.skepticalscience.

com/history-climate-science.html.

Skeptical Science is a non-profit science education program created by John Cook at the University of Queensland. This post on that organization's blog offers a history of climate science which, in chronological order, traces developments from the early 19th century forward to the early 21st century.

Meyer, Robinson. "A Reader's Guide to the Paris Agreement." *Atlantic Monthly, December 2015.* Accessed January 26, 2017. http://www.theatlantic.com/science/archive/2015/12/a-readers-guide-to-the-paris-agreement/420345/.

NASA Earth Observatory. "World of Change -- Global Temperatures." Accessed January 30, 2017. http://earthobservatory.nasa.gov/Features/WorldOfChange/decadaltemp.php.

Nordhaus, William. *The Climate Casino – Risk, Uncertainty and Economics for a Warming World.* New Haven, CT: Yale University Press, 2013.

Nordhaus (1941 -) is a professor of economics at Yale who has studied and written extensively about global warming. The 2°C limit which is now embodied in the 2015 Paris Agreement has been traced back to a paper that he presented in 1976 at the Annual Meeting of the American Economic Association (that paper is accessible at http://www.jstor.org/stable/1815926?seq=1#page_scan_tab_contents). Nordhaus approaches climate change as an intersection point for science, economics and politics. His bottom line is revealed on the opening page: "We are rolling the climatic dice, the outcome will produce surprises, and some of them are likely to be perilous. But we have just entered the Climate Casino, and there is time to turn around and walk out."

Paris Agreement. English version. Accessed February 6, 2017. http://unfccc.int/files/meetings/paris_nov_2015/application/pdf/paris_agreement_english_.pdf

Romm, Joseph. *Climate Change – What Everyone Needs to Know.* New York, NY: Oxford University Press 2016.

Romm (1960 -) has a doctoral degree in physics from MIT and over the past 30 years has held various positions in both the federal government and advisory and advocacy organizations, with a focus upon climate science. He is an advocate for actions to reduce ongoing global warming. Romm has authored and co-authored several books and published papers on the topic, while also serving as science advisor for documentary film productions and providing testimony at Congressional hearings. This book *Climate Change* has been justifiably described to be a primer on the subject; its 268 pages can be recommended to a reader who is looking for a single source that offers a current overview. Chapter 7 offers strategies for our individual lives to reduce carbon emissions.

United Nations Framework Convention on Climate Change. "Fast facts & figures: On the Bali Road Map and Cancun Agreement." Accessed January 29, 2017. http://unfccc.int/essential_background/basic_facts_figures/items/6246.php.

—. "Kyoto Protocol." Accessed February 6, 2017. http://unfccc.int/kyoto_protocol/items/2830.php

—. "Lima Call for Climate Action Puts World on Track to Paris 2015." Accessed January 30, 2017. http://newsroom.unfccc.int/lima/lima-call-for-climate-action-puts-world-on-track-to-paris-2015/.

—. "Lima Climate Change Conference – December 2014." Accessed January 30, 2017. http://unfccc.int/meetings/lima_dec_2014/meeting/8141.php.

—. Text of United Nations Framework Convention on Climate Change 1992, Article 2, page 9. Accessed January 29, 2017. http://unfccc.int/files/essential_background/background_publications_htmlpdf/application/pdf/conveng.pdf.

Vidal, John, Allegra Stratton and Suzanne Goldenburg. "Low targets, goals dropped: Copenhagen ends in failure." Accessed January 29, 2017. *theguardian*, December 18, 2009, https://www.theguardian.com/environment/2009/dec/18/copenhagen-deal.

Weart, Spencer R. *The Discovery of Global Warming*. rev. and expanded ed. Cambridge, MA and London: Harvard University Press. 2008. Supplemented online version January 2017: http://history.aip.org/climate/index.htm#L000.

> Weart (1942 -), before his retirement, was for many years the director of the Center for History of Physics of the American Institute of Physics. He has produced numerous historical articles and three books, including *The Discovery of Global Warming*. He has also taught courses on the history of

science and has given presentations to various audiences. In *The Discovery of Global Warming*, Weart has used both his background in physics and his career in learning and teaching about the history of science to tell how climate science has unfolded to become an increasingly important field. This evolution has coincided with, and in no small way has been facilitated by, the advent and rapid evolution of computers and the modeling they have enabled.

Wikipedia. "Environmental policy in China." Accessed January 29, 2017. https://en.wikipedia.org/wiki/Environmental_policy_in_China.

——. "Kyoto Protocol." Accessed February 7, 2017. https://en.wikipedia.org/wiki/Kyoto_Protocol#Non-ratification_by_the_USA.

Wuebbles, D.J., D.W. Fahey, K.A. Hibbard, D.J. Dokken, B.C. Stewart, and T.K. Maycock, eds., *Highlights* and *Executive Summary, Climate Science Special Report: Fourth National Climate Assessment, Volume I, U.S. Global Change Research Program*, published 2017, accessed November 8, 2017, https://science2017.globalchange.gov/.

This report, in its full 470 pages, is the first of two volumes of the climate assessment mandated by Congress in the Global Change Research Act of 1990. Its preparation was a collaborative effort of several federal agencies, along with selected climate scientists outside the federal government. It presents itself to be a report about the "state of science relating to climate change and its physical impacts" and states that it is "generally intended for those who have a technical background in climate science." The Notes in this book are based primarily upon the Executive Summary in the report; details are to be found in the 15 chapters and five appendices which follow that Executive Summary.

ILLUSTRATIONS

Chapter One – Departure

1.1   Flight 1230 flight path – Source: The author and his brother.

Chapter Two – Numbers

2.1   Knotted strings – Source: Inca Glossary - Appendix C. www.incaglossary.org/appc.html.

2.2   Counting board. – Source: Wikipedia contributors, "counting rods." *Wikipedia. The Free Encyclopedia,* https://en.wikipedia.org/wiki/Counting_rods#/media/File:Checker_counting_board.jpg.

2.3   Abacus – Source: Google Images, "abacus black and white." https://www.google.com/search?q=abacus+black+and+white&tbm=isch&tbs=simg:CAQSlgEJqxVgNR4K5PgaigEL EKjU2AQaBAgDCAoMCxCwjKcIGmEKXwgDEifECJc-DrwicA64IqghbmAOwCPIDqCizNconiynQNLU0rCi-Nu-w1wjYaMIc3waUyfSemKfdAWA2X3kwdqJPystowplNx-1uuppKnDftFo9JE616cOu1VlQ5H-SCAEDAsQjq7-CBoK CggIARIE_1bxUHgw&sa=X&ved=0ahUKEwivyZHX07X SAhUrhlQKHTyRC2sQwg4IGSgA&biw=1672&bih=867.

Chapter Three – Measurements

3.1 Thomas Corwin Mendenhall – Source: Wikipedia contributors, "Thomas Corwin Mendenhall," *Wikipedia, The Free Encyclopedia,* https://en.wikipedia.org/w/index.php?title=Thomas_Corwin_Mendenhall&oldid=766385042.

Chapter Four – Calendars

4.1 Chinese Calendar – Source: Google Images, linked to https://previews.123rf.com/images/jelen80/jelen800902/jelen80090200043/4332147-Chinese-horoscope-Stock-Vector-chinese-zodiac-dragon.jpg.

4.2 International Date Line – Source: Google Images, linked to https://phys.libretexts.org/TextMaps/Astronomy_and_Cosmology_Textmaps/Map%3A_Astronomy_(OpenStax)/4%3A_Earth,_Moon,_and_Sky/4.3%3A_Keeping_Time.

Chapter Five – Clocks

5.1 Su Song Astronomical Clock Tower – Source: David S. Landes, *Revolution in Time – Clocks and the Making of the Modern World,* rev. ed. 2000. Cambridge, MA: Harvard University Press 2000; Landes' source: "Courtesy of John Compton." Image downloaded from Google Images.

5.2 Schematic of mechanical clock – Source: G. J. Whitrow, *Time in History – Views of Time from Prehistory to the Present Day.* New York, NY: Barnes & Noble, Inc. 1988 at p. 103.

Illustrations

5.3 Harrison's first marine chronometer – Source: David S. Landes, *Revolution in Time – Clocks and the Making of the Modern World*, rev. ed. 2000. Cambridge, MA: Harvard University Press 2000; Landes' source: "Courtesy of National Maritime Museum, London".
5.4 Caesium atomic clock, c. 1965 – Source: Internet – Science Photo Library.

Chapter Six – Temperature

6.1 Philo's "thermometer" – Source: Google Images, search "Philo Greek".
6.2 Daniel Gabriel Fahrenheit – Source: https://www.tmatlantic.com/encyclopedia/index.php?ELEMENT_ID=7508
6.3 Anders Celsius – Source: Google Images, https://google.com/search?hl=en&tbm=isch&sxsrf=ACYBGNQm279LaASQQn8au89I7H_KZ1Oteg%3A1581273248561&source=hp&biw=1920&bih=920&ei=oFBAXqGqHZWGtQaz2ouACA&q=anders+celsius+images&oq=Anders+Celsius&gs_l=img.1.0.35i39l2j0l8.4482.10088..14562...1.0..0.62.757.14......0....1..gws-wiz-img....... 0i131.HanHQq_c3LY#imgrc=ZC40d916029fTM&imgdii=2OPwUvrXR7mU3M
6.4 Lord Kelvin (William Thomson) – Source: http://apps.usd.edu/esci/creation/age/content/failed_scientific_clocks/kelvin_cooling.html.

Chapter Seven – Climate

7.1 Greenhouse effect – The author modified diagram from internet.

243

Changing Climate Changing Lives

7.2 The grandfathers of climate science:

Jean Batiste Fourier – Source: *Wikipedia, The Free Encyclopedia*, https://mwl.wikipedia.org/wiki/Jean-Baptiste_Joseph_Fourier#/media/Fexeiro:Joseph_Fourier.jpg; The University of Adelaide.

John Tyndall – Source: Smithsonian Institution Libraries Digital Collection, http://www.sil.si.edu/DigitalCollections/hst/scientific-identity/CF/by_name_display_results.cfm?scientist=Tyndall,%20John.

Svante Arrenhaus – Source: http://www.chemistryexplained.com/A-Ar/Arrhenius-Svante.html.

7.3 Guy Stewart Callendar – Source: *Wikipedia, The Free Encyclopedia,* https://en.wikipedia.org/wiki/Guy_Stewart_Callendar&oldid=9211995124. The University of East Anglia Archive.

7.4 Great Ocean Conveyor Belt – Source: NASA Earth Observatory, https://en.wikipedia.org/wiki/Thermohaline_circulation#/media/File:Thermohaline_Circulation_2.png.

7.5 Carbon dioxide in the Earth's atmosphere – Source: Scripps Institution of Oceanography, https://scripps.ucsd.edu/programs/keelingcurve/2019/06/04/animation-of-keeling-curve-history-updated-to-include-2019-milestone/; https://library.ucsd.edu/dc/collection/bb3381541w.

ACKNOWLEDGMENTS

The seed from which my longstanding interest in the history of science has grown was planted in 1962 by Professor Charles Coulston Gillispie, when I was an undergraduate student in his History of Science course at Princeton. The core reading in that course was his book *The Edge of Objectivity – An Essay in the History of Scientific Ideas*, published in 1960. My marked copy has held a prominent spot on my bookshelf for several decades. In the classroom, Professor Gillispie demonstrated a genuine enthusiasm about the subject of the course, and his enthusiasm was contagious. He was the Dayton-Stockton Professor of History at Princeton and the recipient of numerous professional honors. I remember him as a gentleman-scholar.

I owe many "thank you's" to the many authors of the sources I have found to educate myself about the topics in this book. Those people are the scholars who did the really hard work, and I have enjoyed trying to absorb their knowledge. The challenge for me has been to pick and choose from their writings what I should pass along in my chapters. You, my readers, will judge how well I have met that challenge.

During my writing of this book I have received support from longtime friends. Gary Greer, one of my law partners for about 25 years, applied his inquisitive mind and well-honed editing skills to two drafts. He made many helpful suggestions and raised many questions, several of which sent me searching for answers. Richard and Heather Jacoby also read a draft. They are experienced sailors and celestial navigators who have helped me understand the history of navigation. Richard has also sent me other useful comments. Even so, any mistakes which might yet be found belong to me.

And that brings me to my wife Karen. We were good friends during our high school years, and then things became serious. I have happily shared more than 56 years of my life with her. And she has shared much with me, including her graceful tolerance of my frequent distractions during the preparation of this book.

David R. Johnson